GRAPHING CALCULATOR INSTRUCTION GUIDE

TO ACCOMPANY

CALCULUS CONCEPTS: AN INFORMAL APPROACH TO THE MATHEMATICS OF CHANGE
SECOND EDITION
LaTorre/Kenelly/Fetta/Harris/Carpenter

Iris Brann Fetta
Clemson University

Copyright © 2002 by Houghton Mifflin Company. All rights reserved.

No part of this work may be reproduced or transmitted in any form or by any means, electronic or mechanical, including photocopying and recording, or by any information storage or retrieval system without the prior written permission of Houghton Mifflin Company unless such copying is expressly permitted by federal copyright law. Address inquiries to College Permissions, Houghton Mifflin Company, 222 Berkeley Street, Boston, MA 02116-3764.

Printed in the U.S.A.

ISBN: 0-618-12179-X

2 3 4 5 6 7 8 9 – PO – 05 04 03 02

HOUGHTON MIFFLIN COMPANY   BOSTON   NEW YORK

Preface

This *Guide* uses actual examples and applications in the applied calculus text *Calculus Concepts: An Informal Approach to the Mathematics of Change* by Donald R. LaTorre, John W. Kenelly, Iris B. Fetta, Cynthia R. Harris, and Laurel L. Carpenter. Wherever the technology icon appears in the text, a new technology technique or an effective way to use your calculator for that particular text material is illustrated in this *Guide*. The technology icon is your clue to refer to the particular section of this *Guide* for your specific calculator that discusses the text section. When there is no reference to a certain section in the text, either there is no new procedure to learn or the necessary techniques have been covered in an earlier section of this *Guide*. Refer to the table of contents to find specific types of applications and the *TroubleShooting Guides* in this *Guide* when you have problems.

The use of technology is an integral part of your study of calculus using *Calculus Concepts*. Even though the text requires technology, it does not demand any particular technology. Any graphing calculator or computer algebra system that has the functionality indicated in this *Guide* will suffice. The materials contained herein provide instruction for using the TI-83/TI-83+, or TI-86 in your course. A similar guide for using the TI-89 with the Second Edition of *Calculus Concepts* is available at the *Calculus Concepts* Website. First edition *Graphing Calculator Instruction Guides* for the TI-82, TI-85, and HP-48G/GX graphing calculators are also available at the web site.

This *Guide* is in two parts: Part *A* contains TI-83 instruction and part *B* contains TI-86 materials. Within each part, the discussions are ordered to match the organization of the text chapters. These notation conventions are used to help you recognize various commands and keystrokes:

- Main keyboard keys are enclosed in rectangular boxes except for numeric keys and English alphabet letters; for example, ENTER.

- The second function of a key is listed in parentheses after the main keyboard keystrokes used to activate the second function; for example, 2nd MODE (QUIT).

- The alpha function of a key is listed in parentheses after the main keyboard keystrokes used to activate the alpha function; for example, ALPHA 4 (T).

- Function keys and menu items are indicated by the main keyboard key followed by the keystroke sequence necessary to access the item and then the name of the item in brackets; for example, STAT 1 [EDIT].

This *Guide* does not replace your calculator's instruction manual. You should refer to that manual to learn about the basic operation and use of your calculator. Calculator programs referenced in these materials (for both PC and Macintosh downloads) and *Program Appendices* containing the calculator code for the programs are at the *Calculus Concepts* Website accessible through **college.hmco.com.**

I would like to thank Marika Hoe and Paul Murphy at Houghton Mifflin Company for their help and patience with this project. Any comments or suggestions concerning this *Guide* can be directed to the publisher.

This work is dedicated to the memory of my parents, Iris Burns Brann and Charles Harry Brann.

Iris B. Fetta
Clemson University

Contents

Texas Instruments TI-83/83+ Graphing Calculators .... Part A
Texas Instruments TI-86 Graphing Calculator .... Part B

	A	B
SETUP PAGE	A-1	B-1

BASIC OPERATION
1. Calculating A-2   B-2
2. Using the ANS Memory A-2   B-2
3. Answer Display A-2   B-2
4. Storing Values A-3   B-3
5. Error Messages A-3   B-3

CHAPTER

1 INGREDIENTS OF CHANGE: FUNCTIONS AND LINEAR MODELS

   1.1.1 Entering an Equation in the Graphing List A-4   B-4
   1.1.2 Drawing a Graph A-4   B-4
   1.1.3 Manually Changing the View of a Graph A-5   B-5
          Evaluating Outputs on the Graphics Screen................................... A-5   B-5
   1.1.4 Tracing to Estimate Outputs.. A-6   B-6
   1.1.5 Evaluating Outputs on the Home Screen A-6   B-6
   1.1.6 Evaluating Outputs Using the Table A-7   B-7
   1.1.7 Finding Input Values Using the Solver A-8   B-8
   1.1.8 How to Determine a Guess to Use in the Solver A-9   B-9
   1.1.9 Graphically Finding Intercepts A-10   B-10
          x-Intercept Method A-10   B-11
          Intersection Method A-11   B-12
   1.1.10 Summary of Estimating and Solving Methods A-12   B-12
   1.3.1 Finding the Sum, Difference, Product, Quotient or Composite Function.... A-12   B-13
   1.3.2 Finding a Difference Function A-12   B-13
   1.3.3 Finding a Product Function A-13   B-13
   1.3.4 Checking Your Answer for a Composite Function A-13   B-14
   1.3.5 Turning Functions in the Graphing List Off and On A-14   B-15
   1.3.6 Graphing an Inverse Function A-15   B-15
   1.3.7 Composition Property of Inverse Functions A-16   B-16
   1.3.8 Graphing a Piecewise Continuous Function A-16   B-17
   1.4.1 Estimating Limits Numerically A-18   B-18
          Rounding Off... A-18   B-19
          Rule of Thumb for Determining Limits From Tables.................... A-19   B-19
   1.4.2 Confirming Limits Graphically – Zooming In and Out A-19   B-20
   1.4.3 Investigating End Behavior Numerically A-21   B-21
   1.4.4 Investigating End Behavior Graphically A-22   B-22
   1.5.1 Entering Data... A-22   B-22
   1.5.2 Editing Data A-23   B-23
   1.5.3 Deleting Old Data.. A-23   B-23
   1.5.4 Finding First Differences A-23   B-24

		A-24	B-25
1.5.5	Scatter Plot Setup	A-24	B-25
1.5.6	Drawing a Scatter Plot of Data	A-26	B-25
1.5.7	Finding a Linear Function to Model Data	A-26	B-26
	Graphing the Line of Best Fit	A-27	B-27
1.5.8	Using a Model for Predictions	A-28	B-27
1.5.9	Copying a Graph to Paper	A-28	B-27
1.5.10	Aligning Data	A-28	B-28
1.5.11	Naming and Storing Data (optional)	A-30	B-29
1.5.12	Recalling Stored Data (optional)	A-30	B-30
1.5.13	Deleting Data	A-31	B-30
1.5.14	What is "Best Fit?" (optional)	A-31	B-30

2 INGREDIENTS OF CHANGE: NONLINEAR MODELS

2.1.1	Entering Evenly-Spaced Input Values	A-33	B-32
2.1.2	Finding Percentage Differences	A-33	B-32
2.1.3	Finding an Exponential Function to Model Data	A-34	B-33
2.1.4	Finding a Log Function to Model Data	A-34	B-33
2.1.5	Revisiting Inverse Functions	A-35	B-34
2.2.1	Nominal and Effective Interest Rates	A-36	B-35
2.2.2	More on Solving for Input Values	A-36	B-35
2.2.3	Finding Formula Outputs by Using Previous Entries	A-37	B-36
2.2.4	Determining Future Value	A-38	B-37
2.2.5	Finding Present Value	A-38	B-37
2.3.1	Generating Random Numbers	A-39	B-38
2.3.2	Finding a Logistic Function to Model Data	A-40	B-38
2.3.3	Recalling Model Parameters	A-41	B-40
2.4.1	Finding Second Differences	A-41	B-40
2.4.2	Finding a Quadratic Function to Model Data	A-42	B-41
2.4.3	Finding a Cubic Function to Model Data	A-42	B-41
2.5.1	Shifting Data Vertically	A-44	B-43

3 DESCRIBING CHANGE: RATES

3.1.1	Finding Average Rate of Change	A-45	B-44
3.1.2	Calculating Percentage Change	A-46	B-44
3.2.1	Magnifying a Portion of a Graph	A-46	B-45
3.2.2	Drawing a Tangent Line	A-47	B-46
3.2.3	Visualizing the Limiting Process (optional)	A-47	B-46
3.2.4	Tangent Lines and Instantaneous Rates of Change	A-48	B-47
3.3.1	Calculating Percentage Rate of Change	A-50	B-50

4 DETERMINING CHANGE: DERIVATIVES

4.1.1	Numerically Estimating Slopes on the Home Screen	A-51	B-51
4.1.2	Numerically Estimating Slopes Using the Table	A-52	B-52
4.2.1	Understanding Your Calculator's Slope Function	A-53	B-53
4.4.1	Derivative Notation and Calculator Notation	A-55	B-55
4.4.2	Calculating dy/dx at Specific Input Values	A-56	B-56
4.4.3	Numerically Checking Slope Formulas	A-57	B-57
4.4.4	Graphically Checking Slope Formulas	A-57	B-57
4.5.1	Summary of Checking Methods	A-58	B-58

5 ANALYZING CHANGE: APPLICATIONS OF DERIVATIVES

5.2.1	Finding x-Intercepts of Slope Graphs	A-59	B-59
5.2.2	Finding Zeros of Slope Functions Using the Solver	A-60	B-60
5.2.3	Using the Calculator to Find Optimal Points	A-61	B-61
5.4.1	Finding x-Intercepts of a Second-Derivative Graph	A-62	B-62
5.4.2	Using the Calculator to Find Inflection Points	A-64	B-63

6 ACCUMULATING CHANGE: LIMITS OF SUMS AND THE DEFINITE INTEGRAL

6.1.1	Area Approximations Using Left Rectangles	A-65	B-65
6.1.2	Area Approximations Using Right Rectangles	A-66	B-66
6.1.3	Area Approximations Using Midpoint Rectangles	A-67	B-67
6.1.4	Simplifying Area Approximations	A-68	B-68
6.2.1	Limit of Sums	A-69	B-69
6.4.1	Definite Integral Notation and Calculator Notation	A-70	B-70
6.4.2	The Fundamental Theorem of Calculus	A-71	B-71
6.4.3	Drawing Antiderivative Graphs	A-72	B-72
6.5.1	Evaluating a Definite Integral on the Home Screen	A-73	B-73
6.5.2	Evaluating a Definite Integral from the Graphics Screen	A-74	B-74

7 ANALYZING ACCUMULATED CHANGE: INTEGRALS IN ACTION

7.1.1	Finding the Area between Two Curves	A-75	B-75
7.2.1	Determining the Flow-Rate Function for an Income Stream	A-76	B-76
7.2.2	Future Value of a Discrete Income Stream	A-78	B-78
7.3.1	Consumer Economics	A-80	B-79
7.3.2	Producer Economics	A-81	B-81
7.4.1	Average Value of a Function	A-82	B-82
7.4.2	Geometric Interpretation of Average Value	A-82	B-82
7.5.1	Normal Probabilities	A-83	B-83
7.5.2	Viewing Normal Probabilities	A-83	B-83

8 REPETITIVE CHANGE: CYCLES AND TRIGONOMETRY

8.1.1	Finding Outputs of Trig Functions with Radian Inputs	A-84	B-84
8.2.1	Fitting a Sine Model to Data	A-84	B-84
8.3.1	Derivatives of Sine and Cosine Functions	A-85	B-85
8.5.1	Integrals of Sine and Cosine Functions	A-86	B-86

9 INGREDIENTS OF MULTIVARIABLE CHANGE: MODELS, GRAPHS, RATES

9.1.1	Finding a Cross-Sectional Model from Data (holding the first input variable constant)	A-87	B-87
9.1.2	Finding a Cross-Sectional Model from Data (holding the second input variable constant)	A-88	B-88
9.1.3	Evaluating Outputs of Multivariable Functions	A-88	B-88
9.2.1	Sketching Contour Curves	A-90	B-90
9.3.1	Numerically Checking Partial Derivative Formulas	A-91	B-91
9.3.2	Visualizing and Estimating Partial Rates of Change	A-92	B-92
9.3.3	Finding Partial Rates of Change Using Cross-Sectional Models	A-93	B-93
9.4.1	Evaluating Partial Derivatives of Multivariable Functions	A-94	B-94
9.4.2	Finding the Slope of a Line Tangent to a Contour Curve	A-95	B-95
9.4.3	Compensating for Change	A-95	B-95

10 ANALYZING MULTIVARIABLE CHANGE: OPTIMIZATION

10.2.1 Finding Critical Points Using Matrices A-96	B-96
Using a Matrix of Coefficients A-96	...
Using a Single Matrix A-98	...
10.2.2 Finding Critical Points Using Algebra and the Solver A-99	B-97
10.2.3 Classifying Critical Points Using the Determinant Test A-100	B-99
10.3.1 Finding Optimal Points Under Constrained Optimization A-101	B-100
10.3.2 Classifying Optimal Points Under Constrained Optimization A-102	B-100

11 DYNAMICS OF CHANGE: DIFFERENTIAL EQUATIONS AND PROPORTIONALITY

11.3.1 Euler's Method for a Differential Equation with One Input Variable A-103   B-102
11.3.2 Euler's Method for a Differential Equation with Two Input Variables ... A-104   B-103

TRIGONOMETRY APPENDIX

A.1 Finding Outputs of Trig Functions with Degree Inputs A-105   B-104
A.2 Converting Angles From Radians to Degrees A-106   B-105
A.2 Converting Angles From Degrees to Radians A-106   B-105

TROUBLESHOOTING GUIDE A-107   B-106

PROGRAM APPENDIX   At the *Calculus Concepts* Website

The *TI-89 Graphing Calculator Instruction Guide* for the 2nd Edition of the text is available at the *Calculus Concepts* Website.

Note that 1st Edition *Graphing Calculator Instruction Guides* for the TI-82, TI-85, and HP-48G/GX calculators are also available at the *Calculus Concepts* Website.

How to Use This Guide

First, go to either part *A* or part *B* of this *Guide*, depending on which model calculator you are using. Then, find the particular section in this *Guide* that references the technology icon in the text, marked with [icon], according to the table of contents section number. Section numbers in the contents should be read as follows:

> Section $x.y.z$ of this *Guide* refers to the discussion in Chapter x, Section y of the *Calculus Concepts* text. The discussion is the zth topic in this *Guide* that refers to that particular chapter and section.

For example, the detailed contents indicate that there are 8 different sections in the *Guide* (Section 1.3.1 through Section 1.3.8) pertaining to Section 1.3 in *Calculus Concepts*. The discussion in Section 1.3.2 of this *Guide* is the second technology discussion pertaining to Chapter 1, Section 1.3 of your text.

Part A

Guide for Texas Instruments TI-83 Graphing Calculator

This *Guide* is designed to offer step-by-step instruction for using your TI-83 or TI-83 Plus graphing calculator with the text *Calculus Concepts: An Informal Approach to the Mathematics of Change*. A technology icon next to a particular example or discussion in the text directs you to a specific portion of this *Guide*. You should also utilize the table of contents in this *Guide* to find specific topics on which you need instruction.

Setup Instructions

Before you begin, check the TI-83 setup and make sure the settings described below are chosen. Whenever you use this *Guide*, we assume (unless instructed otherwise) that your calculator settings are as shown in Figures 1, 2, and 3.

- Press MODE and choose the settings shown in Figure 1 for the basic setup.

- Specify the statistical setup with STAT 5 [SetUpEditor] followed by 2nd 1 ⎕, 2nd 2 ⎕, 2nd 3 ⎕, 2nd 4 ⎕, 2nd 5 ⎕, and 2nd 6. Press ENTER to view the screen in Figure 2.

- Check the window format by pressing 2nd ZOOM (FORMAT) and choose the settings shown in Figure 3.

  - If you do not have the darkened choices shown in Figure 1 and Figure 3, use the arrow keys to move the blinking cursor over the setting you want to choose and press ENTER.

  - Press 2nd MODE (QUIT) to return to the home screen.

TI-83 Basic Setup
Figure 1

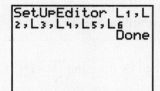
TI-83 Statistical Setup
Figure 2

TI-83 Window Setup
Figure 3

Basic Operation

You should be familiar with the basic operation of your calculator. With your calculator in hand, go through each of the following.

1. **CALCULATING** You can type in lengthy expressions; just make sure that you use parentheses when you are not sure of the calculator's order of operations. Always enclose in parentheses any numerators and denominators of fractions and powers that consist of more than one term.

   Evaluate $\dfrac{1}{4*15+\dfrac{895}{7}}$. Enclose the denominator in parentheses so that the addition is performed before the division into 1. It is not necessary to use parentheses around the fraction 895/7.

   Evaluate $\dfrac{(-3)^4-5}{8+1.456}$. Use $\boxed{(-)}$ for the negative symbol and $\boxed{-}$ for the subtraction sign. To clear the home screen, press $\boxed{\text{CLEAR}}$.

   ```
   1/(4*15+895/7)
              .0053231939
   ((-3)^4-5)/(8+1.
   456)
              8.037225042
   ```

   NOTE: The numerator and denominator must be enclosed in parentheses and $-3^4 \neq (-3)^4$.

   Now, evaluate e^3*2. The TI-83 prints the left parenthesis when you press $\boxed{\text{2nd}}$ $\boxed{\text{LN}}$ (e^x). The calculator assumes that you are inserting a right parenthesis at the end of what you type unless you insert one elsewhere. Also, if you press $\boxed{(}$ after pressing the key for e and do not include another right parenthesis, the TI-83 assumes that you are including everything following the extra parenthesis in the exponent.

   ```
   e^(3)*2
              40.17107385
   e^((3)*2
              403.4287935
   e^(3*2
              403.4287935
   ```

2. **USING THE ANS MEMORY** Instead of again typing an expression that was just evaluated, use the answer memory by pressing $\boxed{\text{2nd}}$ $\boxed{(-)}$ (ANS).

   Calculate $\left(\dfrac{1}{4*15+\dfrac{895}{7}}\right)^{-1}$ using this nice shortcut.

   Type Ans^{-1} by pressing $\boxed{\text{2nd}}$ $\boxed{(-)}$ (ANS) $\boxed{x^{-1}}$.

   ```
   895/7
              127.8571429
   1/(4*15+Ans)
              .0053231939
   Ans-1
              187.8571429
   ```

3. **ANSWER DISPLAY** When the denominator of a fraction has no more than three digits, your calculator can provide the answer in the form of a fraction. When an answer is very large or very small, the calculator displays the result in scientific notation.

   The "to a fraction" key is obtained by pressing $\boxed{\text{MATH}}$ 1 [▶Frac].

   ```
   2/5+1/3
              .7333333333
   Ans▶Frac
                   11/15
   .3875▶Frac
                   31/80
   ```

The calculator's symbol for *times* 10^{12} is E12. Thus, 7.945E12 means 7,945,000,000,000.

The result 1.4675E−6 means $1.4675*10^{-6}$, which is the scientific notation expression for 0.0000014675.

4. **STORING VALUES** Sometimes it is beneficial to store numbers or expressions for later recall. To store a number, type the number, press STO▶ ALPHA, type the letter corresponding to the storage location, and then press ENTER. To join several short commands together, use ALPHA . (:) between the statements. Note that when you join statements with a colon, only the value of the last statement is shown.

Store 5 in *A* and 3 in *B*, and then calculate $4A - 2B$.

To recall a stored value, press ALPHA, type the letter in which the expression or value is stored, and then press ENTER.

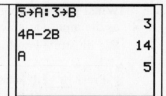

- There are 27 memory locations in the TI-83: A through Z and θ. Whatever you store in a particular memory location stays there until it is replaced by something else. Before you use a particular location, 0 is the stored value.

5. **ERROR MESSAGES** When your input is incorrect, the TI-83 displays an error message.

If you have more than one command on a line without the commands separated by a colon (:), an error message results when you press ENTER.

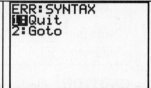

- When you get an error message, Choose 2 [Goto] to position the cursor to where the error occurred so that you can correct the mistake or choose 1 [Quit] to begin a new line on the home screen.

- A common mistake is using the negative symbol (−) instead of the subtraction sign − or vice-versa.

If you try to store something to a particular memory location that is being used for a different type of object, a DATA TYPE error results. Consult either *TroubleShooting the TI-83* in this *Guide* or your *TI-83 Owner's Guidebook*.

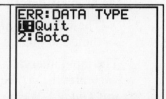

- When you are executing a program that you have transferred to or input in your calculator, you should always choose the 1: Quit option. Choosing 2: Goto will call up the program code, and you may inadvertently change the program so that it will not properly execute.

- Other error messages are referred to throughout this *Guide* at the points that they might occur in specific applications.

Chapter 1 Ingredients of Change: Functions and Linear Models

 ## 1.1 Models, Functions, and Graphs

There are many uses for a function that is entered in the calculator's graphing list. Graphing the function in an appropriate viewing window is one of these. Because you must enter all functions on one line (that is, you cannot write fractions and exponents the same way you do on paper), it is very important to have a good understanding of the calculator's order of operations and to use parentheses whenever they are needed.

1.1.1 ENTERING AN EQUATION IN THE GRAPHING LIST The graphing list contains space for 10 equations, and the output variables are called by the names Y1, Y2, ..., and Y0. When you want to graph an equation, you should enter in the graphing list. You must use X as the input variable if you intend to draw the graph of the equation or use the TI-83 table. We illustrate graphing with the equation in Example 4 of Section 1.1: $v(t) = 3.622(1.093^t)$.

Press $\boxed{Y=}$ to access the graphing list. If there are any previously entered equations that you will no longer use, delete them from the graphing list.	To delete an equation from the Y= list, position the cursor on the line with the equation. Press $\boxed{CLEAR}$.
| For convenience, we use the first, or Y1, location in the list. We intend to graph this equation, so the input variable must be called *x*, not *t*. Enter the right-hand side of the equation, $3.622(1.093^x)$, with 3 $\boxed{.}$ 6 2 2 $\boxed{(}$ 1 $\boxed{.}$ 0 9 3 $\boxed{\wedge}$ $\boxed{X,T,\theta,n}$ $\boxed{)}$ <br><br> Note that you should use the $\boxed{X,T,\theta,n}$ key for *x*, not the times sign key, $\boxed{\times}$. | ```
Plot1 Plot2 Plot3
\Y1■3.622(1.093^
X)
\Y2=
\Y3=
\Y4=
\Y5=
\Y6=
``` |

CAUTION: Plot1, Plot2, and Plot3 at the top of the Y= list should *not* be darkened when you are graphing an equation and not graphing data points. If any of these is darkened, use $\boxed{\blacktriangle}$ until you are on the darkened plot name. Press $\boxed{ENTER}$ to make the name(s) not dark. If you do not do this, you may receive a STAT PLOT error message.

1.1.2 DRAWING A GRAPH As is the case with most applied problems in *Calculus Concepts*, the problem description indicates the valid input interval. Consider Example 4 of Section 1.1:

> The value of a piece of property between 1980 and 2000 is given by $v(t) = 3.622(1.093^t)$ thousand dollars where *t* is the number of years since the end of 1980.

The input interval is 1980 ($t = 0$) to 2000 ($t = 20$). Before drawing the graph of *v* on this interval, enter the $v(t)$ equation in the Y= list using *x* as the input variable. (See Section 1.1.1 of this *Guide*.) We now draw the graph of the function *v* for *x* between 0 and 20.

| Press $\boxed{WINDOW}$ to set the view for the graph. Enter 0 for Xmin and 20 for Xmax. (For 10 tick marks between 0 and 20, enter 2 for Xscl. If you want 20 tick marks, enter 1 for Xscl, etc. Xscl does not affect the shape of the graph. Ignore the other numbers – we set their values in the next set of instructions.) | ```
WINDOW
Xmin=0
Xmax=20
Xscl=2
Ymin=-6
Ymax=6
Yscl=1
Xres=1
``` |
|---|---|

Xmin and Xmax are, respectively, the settings of the left and right edges of the viewing screen, and Ymin and Ymax are, respectively, the settings for the lower and upper edges of the viewing screen. Xscl and Yscl set the spacing between the tick marks on the x- and y-axes. (Leave Xres set to 1 for all applications in this *Guide*.)

To have the TI-83 determine the view for the output, press ZOOM ▲ [ZoomFit]. Now press ENTER to see the graph of the function v.	

Note that any vertical line drawn on this graph intersects it in only one point, so the graph does represent a function.

Press WINDOW to see the view set by Zoomfit. The view has $0 \leq x \leq 20$ and $3.622 \leq y \leq 21.446...$. (Note that ZoomFit did not change the x-values that you set manually.)	WINDOW Xmin=0 Xmax=20 Xscl=2 Ymin=3.622 Ymax=21.446265... Yscl=1 Xres=1

1.1.3 MANUALLY CHANGING THE VIEW OF A GRAPH

We just saw how to have the TI-83 set the view for the output variable. Whenever you draw a graph, you can also manually set or change the view for the output variable. If for some reason you do not have an acceptable view of a graph or if you do not see a graph, change the view for the output variable with one of the ZOOM options or manually set the WINDOW until you see a good graph. (We will later discuss other ZOOM options.) We continue using the function v in Example 4 of Section 1.1, but assume here that you have not yet drawn the graph of v.

Press WINDOW, enter 0 for Xmin, 20 for Xmax, and (assuming we do not know what to use for the vertical view), enter some arbitrary values for Ymin and Ymax. (The graph to the right was drawn with Ymin = −5 and Ymax = 5). Press GRAPH.	
Evaluating Outputs on the Graphics Screen: Press TRACE*. Recall we are given in this application that the input variable is between 0 and 20. If you now type the number that you want to substitute into the function whose graph is drawn, say 0, you see the screen to the right. Note that the equation of the function whose graph you are drawing appears at the top of the screen.	Y1=3.622(1.093^X) X=0
Press ENTER and the input value is substituted in the function. The input and output values are shown at the bottom of the screen. (This method works even if you do not see any of the graph on the screen.)	Y1=3.622(1.093^X) X=0 Y=3.622
Substitute the right endpoint of the input interval into the function by pressing 20 ENTER. We see that two points on this function are approximately (0, 3.622) and (20, 21.446).	Y1=3.622(1.093^X) X=20 Y=21.446266

Press WINDOW, enter 3.5 for Ymin and 22 for Ymax, and press GRAPH. If the graph you obtain is not a good view of the function, repeat the above process using *x*-values in between the two endpoints to see if the output range should be extended in either direction. (Note that the choice of the values 3.5 and 22 was arbitrary. Any values close to the outputs in the points you find are also acceptable.)

*Instead of using TRACE, you could use the TI-83 TABLE or evaluate the function at 0 and 20 on the home screen to find the range of values for the output value. We later discuss using these features.

1.1.4 TRACING TO ESTIMATE OUTPUTS
You can display the coordinates of certain points on the graph by tracing. Unlike the substitution feature of TRACE that was discussed on the previous page, the *x*-values that you see when tracing the graph depend on the horizontal view that you choose. The output values that are displayed at the bottom of the screen are calculated by substituting the *x*-values into the equation that is being graphed. We again assume that you have the function $v(x) = 3.622(1.093^x)$ entered in the Y1 location of the Y= list.

With the graph on the screen, press TRACE, press and hold ▶ to move the trace cursor to the right, and press and hold ◀ to move the trace cursor to the left. The equation that you are tracing appears at the top of the graphing screen.

Trace past one edge of the screen and notice that even though you cannot see the trace cursor, the *x*- and *y*-values of points on the line are still displayed at the bottom of the screen. Also note that the graph scrolls to the left or right as you move the cursor past the edge of the current viewing screen.

Use either ▶ or ◀ to move the cursor near *x* = 15. We estimate that *y* is *approximately* 13.9 when *x* is *about* 15.

It is important to realize that trace outputs should *never* be given as answers to a problem unless the displayed *x*-value is *identically* the same as the value of the input variable.

1.1.5 EVALUATING OUTPUTS ON THE HOME SCREEN
The input values used in the evaluation process are *actual* values, not estimated values such as those generally obtained by tracing near a certain value. Actual values are also obtained when you evaluate outputs from the graphing screen using the process that was discussed in Section 1.1.3 of this *Guide*.

We again consider the function $v(t) = 3.622(1.093^t)$ that is in Example 4 of Section 1.1.

Using *x* as the input variable, enter Y1 = 3.622(1.093)^X. Return to the home screen by pressing 2nd MODE (QUIT).

Go to the Y-VARS menu with VARS ▶ [Y-VARS]. Choose 1: Function by pressing the number 1 or ENTER.

NOTE: We choose Y1 as the function location most of the time, but you can use any of the ten available locations. If you do, replace Y1 in the instructions with the function you choose.

TI-83 Guide

Choose 1: Y1 by pressing the number 1 or ENTER. Y1 appears on the home screen. Press (, type the *x*-value to be substituted into the equation, and press). Then press ENTER.	FUNCTION 1:Y1 2:Y2 3:Y3 4:Y4 5:Y5 6:Y6 7↓Y7	Y1(15) 13.74835936

NOTE: You do not have to have the *closing* parentheses on the right if nothing else follows it.

It is now a simple matter to evaluate the function at other inputs. For instance, substitute $x = 20$ into the equation by recalling the previous entry with 2nd ENTER (ENTRY), change 15 to 20 by pressing ◄ ◄ ◄ and typing 20, and then press ENTER.	Y1(15)
	   13.74835936
	Y1(20)
	   21.44626584

Evaluate Y1 at $x = 0$ by recalling the previous entry with 2nd ENTER (ENTRY), change 20 to 0 by pressing ◄ ◄ ◄ DEL, and then press ENTER.	Y1(15)
	   13.74835936
	Y1(20)
	   21.44626584
	Y1(0)
	   3.622

1.1.6 EVALUATING OUTPUTS USING THE TABLE

Function outputs can be determined by evaluating on the graphics screen, as discussed in Section 1.1.3, or by evaluating on the home screen as discussed in Section 1.1.5 of this *Guide*. You can also evaluate functions using the TI-83 TABLE. When you use the table, you can either enter specific input values and find the outputs or generate a list of input and output values in which the inputs begin with TblStart and differ by ΔTbl.

Let's use the TABLE to evaluate the function $v(t) = 3.622(1.093^t)$ at the input $t = 15$. Even though you can use any of the function locations, we again choose to use Y1. Press Y=, clear the function locations, and enter 3.622(1.093)^X in location Y1 of the Y= list.

Choose TABLE SETUP with 2nd WINDOW (TBLSET). To generate a list of values beginning with 13 such that the table values differ by 1, enter 13 in the TblStart location, 1 in the ΔTbl location, and choose AUTO in the Indpnt: and Depend: locations. (Remember that you "choose" a certain setting by moving the cursor to that location and pressing ENTER.)	TABLE SETUP TblStart=13 ΔTbl=1 Indpnt: **Auto** Ask Depend: **Auto** Ask
Press 2nd GRAPH (TABLE), and observe the list of input and output values. Notice that you can scroll through the table with ▼, ▲, ◄, and/or ►. The table values may be rounded in the table display. You can see more of the output by highlighting a particular value and viewing the bottom of the screen.	X | Y1 13 | 11.508 14 | 12.579 15 | 13.748 16 | 15.027 17 | 16.424 18 | 17.952 19 | 19.621 Y1=13.7483593553

If you are interested in evaluating a function at inputs that are not evenly spaced and/or you only need a few outputs, you should use the ASK feature of the table instead of using AUTO. Note that when using ASK, the settings for TblStart and ΔTbl do not matter.

Return to the table set-up with 2nd [WINDOW] (TBLSET), and choose ASK in the Indpnt: location. Press [ENTER]. Press 2nd [GRAPH] (TABLE) to go to the table and enter the *x*-values.	TABLE SETUP TblStart=13 △Tbl=1 Indpnt: Auto **Ask** Depend: **Auto** Ask	X Y1 15 13.748 0 3.622 20 21.446 Y1=13.7483593553

NOTE: Unwanted entries or values from a previous problem can be cleared with [DEL].

1.1.7 FINDING INPUT VALUES USING THE SOLVER Your calculator solves for the input values of any equation that is in the form *"expression = 0"*. This means that all terms must be on one side of the equation and 0 must be on the other side before you enter the equation into the calculator. The expression can, but does not have to, use *x* as the input variable.

The TI-83 offers several methods of solving for input variables. We first illustrate using the SOLVER. (Solving using graphical methods will be discussed after using the SOLVER is explored.) You can refer to an equation that you have already entered in the Y= list or you can enter the equation in the solver.

Return to the home screen with 2nd [MODE] (QUIT). Access the solver by pressing [MATH] 0. If there are no equations stored in the solver, you will see the screen displayed on the right – or	EQUATION SOLVER eqn:0=■
– if the solver has been previously used, you will see a screen similar to the one shown on the right. If this is the case, press ▲ and [CLEAR] to delete the old equation. You should then have the screen shown in the previous step.	2X²−3=0 X=2 bound={-1E99,1...

Let's now use the solver to answer the question in part *e* of Example 4 in Section 1.1: "When did the land value reach \$20,000?" Because the land value is given by $v(t) = 3.622(1.093^t)$ thousand dollars where *t* is the number of years after the end of 1980, we are asked to *solve* the equation $3.622(1.093^t) = 20$. That is, we are asked to find the input value *t* that makes this equation a true statement. We must enter the equation into the solver with a 0 on one side, so subtract 20 from each side of $3.622(1.093^t) = 20$ to obtain the equation $3.622(1.093^t) - 20 = 0$.

If you already have Y1 = 3.622(1.093^X) in the graphing list, you can refer to the function as Y1 in the SOLVER. If not, you can enter 3.622(1.093^X) instead of Y1 in the eqn: location of the SOLVER. Press [ENTER].	EQUATION SOLVER eqn:0=Y1−20
If you need to edit the equation, press ▲ until the previous screen reappears. Edit the equation and then return here. With the cursor in the X location, enter a guess – say 19. (You could have also used as a guess* the value that was in the X location when you accessed the SOLVER.)	Y1−20=0 X=19■ bound={-1E99,1...

*More information on entering a guess appears at the end of this discussion.

CAUTION: You should not change anything in the "bound" location of the SOLVER. The values in that location are the ones between which the TI-83 searches[1] for a solution. If you should accidentally change anything in this location, exit the solver, and begin the entire process again. (The bound is automatically reset when you exit the SOLVER.)

Be certain that the cursor is on the line corresponding to the input variable for which you are solving (in this example, X). Solve for the input by pressing [ALPHA] [ENTER] (SOLVE). The answer to the original question is that the land value was $20,000 about 19.2 years after 1980 – *i.e.*, in the year 2000.	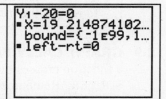

- Notice the black dot that appears next to X and next to the last line on the above screen. This is the TI-83's way of telling you that a solution has been found. The bottom line on the screen that states left – rt = 0 indicates the value found for X is an exact solution since both sides of the equation evaluate to the same quantity.

- If a solution continues beyond the edge of the calculator screen, you see "..." to the right of the value. Be certain that you press and hold [▶] to scroll to the end of the number.

  The value may be given in scientific notation, and the portion that you cannot see determines the location of the decimal point. (See *Basic Operation, #3*, in this *Guide*.)

1.1.8 HOW TO DETERMINE A GUESS TO USE IN THE EQUATION SOLVER What you use in the solver as a guess tells the TI-83 where to start looking for the answer. How close your guess is to the actual answer is not very important unless there is more than one solution to the equation. *If the equation has more than one answer, the solver will return the solution that is closest to the guess you supply.* In such cases, you need to know how many answers you should search for and their approximate locations.

Three of the methods that you can use to estimate the value of a guess for an answer from the SOLVER follow. We illustrate these methods using the land value function from Example 4 of Section 1.1 and the equation $v(t) = 3.622(1.093^t) = 20$.

1. Enter the function in some location of the graphing list – say Y1 = 3.622(1.093^X) and draw a graph of the function. Press [TRACE] and hold down either [▶] or [◀] until you have an estimate of where the output is 20. Use this X-value, 19 or 19.4 or 19.36, as your guess in the SOLVER.	

Remember that you can use any letter to represent the input in the solver, but to graph a function, you must use X as the input variable.

2. Enter the left- and right-hand sides of the equation in two different locations of the Y= list – say Y1 = 3.622(1.093^X) and Y2 = 20. With the graph on the screen, press [TRACE] and hold down either [▶] or [◀] until you get an estimate of the X-value where the curve crosses the horizontal line representing 20.	

---

[1] It is possible to change the bound if the calculator has trouble finding a solution to a particular equation. This, however, should happen rarely. Refer to the *TI-86 Graphing Calculator Guidebook* for details.

3. Use the AUTO setting in the TABLE, and with ▲ or ▼ scroll through the table until a value near the estimated output is found. Use this X-value or a number near it as your guess in the SOLVER. (Refer to Section 1.1.6 of this *Guide* to review the instructions for using the TABLE.)

1.1.9 GRAPHICALLY FINDING INTERCEPTS

Finding the input value at which the graph of a function crosses the vertical and/or horizontal axis can be found graphically or by using the SOLVER. Remember the process by which we find intercepts:

- To find the y-intercept of a function $y = f(x)$, set $x = 0$ and solve the resulting equation.
- To find the x-intercept of a function $y = f(x)$, set $y = 0$ and solve the resulting equation.

An *intercept* is the where the graph crosses or touches an axis. Also remember that the x-intercept of the function $y = f(x)$ has the same value as the root or solution of the equation $f(x) = 0$. **Thus, finding the x-intercept of the graph of $f(x) - c = 0$ is the same as solving the equation $f(x) = c$.**

We illustrate this method with a problem similar to the one in Activity 40 in Section 1.1 of *Calculus Concepts*. You should practice by solving the equation $3.622(1.093^x) = 20$ using this graphical method and by solving the equation that follows using the SOLVER.

Suppose we are asked to find the input value of $f(x) = 3x - 0.8x^2 + 4$ that corresponds to the output $f(x) = 2.3$. That is, we are asked to find x such that $3x - 0.8x^2 + 4 = 2.3$. Because this function is not given in a context, we have no indication of an interval of input values to use when drawing the graph. We will use the ZOOM features to set an initial view and then manually set the WINDOW until we see a graph that shows the important points of the function (in this case, the intercept or intercepts.) You can solve this equation graphically using either the *x-intercept method* or the *intersection method*. We present both, and you should use the one you prefer.

X-INTERCEPT METHOD for solving the equation $f(x) - c = 0$:

Press Y= and clear all locations with CLEAR. Enter the function $3x - 0.8x^2 + 4 - 2.3$ in Y1. You can enter x^2 with X,T,θ,n x^2 or enter it with X,T,θ,n ^ 2. Remember to use −, not (−), for the subtraction signs.	
Draw the graph of Y1 with ZOOM 4 [Zdecimal] or ZOOM 6 [ZStandard]. If you use the former, press WINDOW and reset Xmax to 6 and Ymax to 5 to get a better view the graph. (If you reset the window, press GRAPH to draw the graph.)	
To graphically find an x-intercept, *i.e.*, a value of x at which the graph crosses the horizontal axis, press 2nd TRACE (CALC) 2 [zero]. Press and hold ◄ until you are near, but to the *left* of, the leftmost x-intercept. Press ENTER to mark the location of the *left* bound for the x-intercept.	

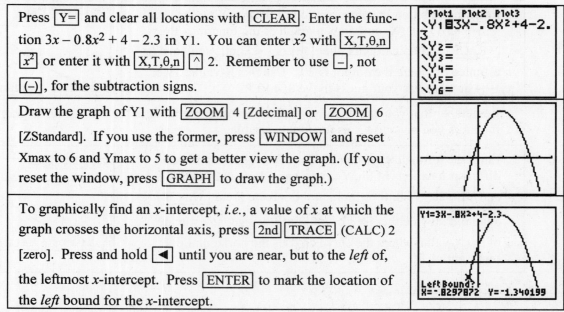

Notice the small arrowhead (▶) that appears above the location to mark the left bound. Now press and hold ▶ until you are to the *right* of this *x*-intercept. Press ENTER to mark the location of the *right* bound for the *x*-intercept.	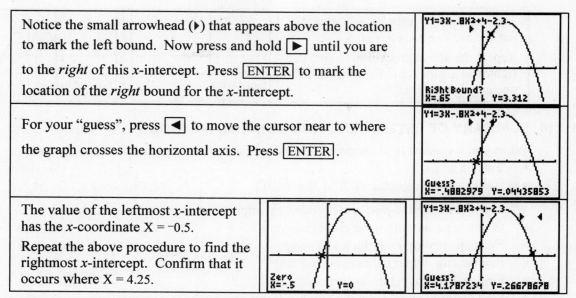
For your "guess", press ◀ to move the cursor near to where the graph crosses the horizontal axis. Press ENTER.	
The value of the leftmost *x*-intercept has the *x*-coordinate X = −0.5. Repeat the above procedure to find the rightmost *x*-intercept. Confirm that it occurs where X = 4.25.	

NOTE: If this process does not return the correct value for the intercept that you are trying to find, you have probably not included the place where the graph crosses the axis between the two bounds (*i.e.*, between the ▶ and ◀ marks on the graph.)

INTERSECTION METHOD for solving the equation $f(x) = c$:

Press Y= and clear all locations with CLEAR. Enter one side of the equation, $3x - 0.8x^2 + 4$ in Y1 and the other side of the equation, 2.3, in Y2.	
Draw the graphs with ZOOM 4 [ZDecimal] or ZOOM 6 [ZStandard]. If you use the former, press WINDOW and reset Xmax to 6 and Ymax to 8 to get a better view the graph. (If you reset the window, press GRAPH to draw the graph.)	
To locate where Y1 = Y2, press 2nd TRACE (CALC) 5 [intersect]. Press ENTER to mark the first curve. The cursor jumps to the other function – here, the line. Next, press ENTER to mark the second curve.	
Note that the equation of each function appears on the top left of the screen as the cursor moves from the curve to the line. Next, supply a guess for the point of intersection. Use ◀ to move the cursor near the intersection point you want to find – in this case, the leftmost point. Press ENTER.	

The value of the leftmost *x*-intercept has the *x*-coordinate X = −0.5.

Repeat the above procedure to find the rightmost *x*-intercept. Confirm that it occurs where X = 4.25.

1.1.10 SUMMARY OF ESTIMATING AND SOLVING METHODS Use the method you prefer.

When you are asked to *estimate* or *approximate* an output or an input value, you can:
- Trace a graph   (Section 1.1.4)
- Use close values obtained from the TABLE   (Sections 1.1.3, 1.1.5, 1.1.7)

When you are asked to *find* or *determine* an output or an input value, you should:
- Evaluate an output on the graphics screen   (Section 1.1.3)
- Evaluate an output on the home screen   (Section 1.1.5)
- Evaluate an output value using the table   (Section 1.1.6)
- Find an input using the solver   (Sections 1.1.7, 1.1.8)
- Find an input value from the graphic screen (using the *x*-intercept method or the intersection method)   (Section 1.1.9)

1.3 Constructed Functions

Your calculator can find output values of and graph combinations of functions in the same way that you do these things for a single function. The only additional information you need is how to enter constructed functions in the graphing list.

1.3.1 FINDING THE SUM, DIFFERENCE, PRODUCT, QUOTIENT OR COMPOSITE FUNCTION
Suppose that a function *f* has been entered in Y1 and a function *g* has been entered in Y2. Your calculator will evaluate and graph these constructed functions:

Enter Y1 + Y2 in Y3 to obtain the sum function $(f + g)(x) = f(x) + g(x)$.
Enter Y1 − Y2 in Y4 to obtain the difference function $(f - g)(x) = f(x) - g(x)$.
Enter Y1*Y2 in Y5 to obtain the product function $(f \cdot g)(x) = f(x) \cdot g(x)$.
Enter Y1/Y2 in Y6 to obtain the quotient function $(f \div g)(x) = \dfrac{f(x)}{g(x)}$.
Enter Y1(Y2) in Y7 to obtain the composite function $(f \circ g)(x) = f(g(x))$.

The functions can be entered in any location in the Y= list. Although the TI-83 will not give an algebraic formula for a constructed function, you can check your final answer by evaluating your constructed function and the calculator-constructed function at several different points to see if they yield the same outputs.

1.3.2 FINDING A DIFFERENCE FUNCTION
We illustrate this technique with the functions that are given on page 31 of Section 1.3 of *Calculus Concepts*: Sales = $S(t) = 3.570(1.105^t)$ million dollars and costs = $C(t) = -39.2t^2 + 540.1t + 1061.0$ thousand dollars *t* years after 1996.

Press [Y=]. Clear previously-entered equations with [CLEAR].

Enter *S* in Y1 by pressing 3 [.] 570 [(] 1 [.] 105 [^] [X,T,θ,n] [)]

[ENTER] and enter *C* in Y2 by pressing [(−)] 39 [.] 2 [X,T,θ,n]

[x²] [+] 540 [.] 1 [X,T,θ,n] [+] 1061 [ENTER].

```
Plot1 Plot2 Plot3
\Y1■3.570(1.105^
X)
\Y2■-39.2X²+540.
1X+1061
\Y3=■
\Y4=
\Y5=
```

TI-83 Guide

The difference function, the profit $P(x) = S(x) - 0.001C(x) =$ Y1 − 0.001Y2, is entered in Y3 with [VARS] [▶] 1 [Function] 1 [Y1] [−] [.] 001 [VARS] [▶] 1 [Function] 2 [Y2].	Plot1 Plot2 Plot3 \Y1■3.570(1.105^X) \Y2■-39.2X²+540.1X+1061 \Y3■Y1−.001Y2 \Y4= \Y5=
To find the profit in 1998, evaluate Y3 when X = 2. Return to the home screen with [2nd] [MODE] (QUIT). Press [VARS] [▶] 1 [Function] 3 [Y3] [(] 2 [)] [ENTER] to see the result. We find that the profit in 1998 was $P(2) \approx 2.375$ million dollars.	Y3(2) 2.37465925

- You can evaluate on the home screen, graphics screen, or in the table. We choose to use the home screen, but you should choose the method you prefer.

1.3.3 FINDING A PRODUCT FUNCTION We illustrate this technique with the functions that are given on page 32 of Section 1.3 of *Calculus Concepts*: Milk price = $S(x) = 0.007x + 1.492$ dollars per gallon on the xth day of last month and milk sales = $G(x) = 31 - 6.332(0.921^x)$ gallons of milk sold on the xth day of last month.

Press [Y=], and clear each previously-entered equation with [CLEAR]. Enter S in Y1 by pressing [.] 007 [X,T,θ,n] [+] 1 [.] 492 [ENTER] and enter G in Y2 by pressing 31 [−] 6 [.] 332 [(] [.] 921 [^] [X,T,θ,n] [)] [ENTER].	Plot1 Plot2 Plot3 \Y1■.007X+1.492 \Y2■31−6.332(.921^X) \Y3=■ \Y4= \Y5= \Y6=
Enter the product function $T(x) = S(x) \cdot G(x) = $ Y1·Y2 in Y3 with the keystrokes [VARS] [▶] 1 [Function] 1 [Y1] [VARS] [▶] 1 [Function] 2 [Y2].	Plot1 Plot2 Plot3 \Y1■.007X+1.492 \Y2■31−6.332(.921^X) \Y3■Y1Y2 \Y4= \Y5= \Y6=

NOTE: You do not have to, but you can, use [×] between Y1 and Y2 to indicate a product function. You cannot use parentheses to indicate the product function because the TI-83 will think that you are entering Y3 as a composite function.

To find milk sales on the 5th day of last month, evaluate Y3 at X = 5. We choose to do this on the home screen. Return to the home screen with [2nd] [MODE] (QUIT). Press [VARS] [▶] 1 [Function] 3 [Y3] [(] 5 [)] [ENTER] to see the result. We find that milk sales were $T(5) \approx \$40.93$.	Y3(5) 40.92965522

1.3.4 CHECKING YOUR ANSWER FOR A COMPOSITE FUNCTION We illustrate this technique with the functions that are given on page 33 of Section 1.3 of *Calculus Concepts*: altitude = $F(t) = -222.22t^3 + 1755.95t^2 + 1680.56t + 4416.67$ feet above sea level where t is the time into flight in minutes and air temperature = $A(F) = 277.897(0.99984^F) - 66$ degrees Fahrenheit where F is the number of feet above sea level. Remember that when you enter functions in the Y= list, you must use X as the input variable.

Press $\boxed{Y=}$, and clear each previously entered equation with $\boxed{\text{CLEAR}}$. Enter F in Y1 by pressing $\boxed{(-)}$ 222 $\boxed{.}$ 22 $\boxed{X,T,\theta,n}$ $\boxed{\wedge}$ 3 $\boxed{+}$ 1755 $\boxed{.}$ 95 $\boxed{X,T,\theta,n}$ $\boxed{x^2}$ $\boxed{+}$ 1680 $\boxed{.}$ 56 $\boxed{X,T,\theta,n}$ $\boxed{+}$ 4416 $\boxed{.}$ 67 $\boxed{\text{ENTER}}$ and enter A in Y2 by pressing 277 $\boxed{.}$ 897 $\boxed{(}$ $\boxed{.}$ 99984 $\boxed{\wedge}$ $\boxed{X,T,\theta,n}$ $\boxed{)}$ − 66 $\boxed{\text{ENTER}}$.

Enter the composite function $(A \circ F)(x) = A(F(x)) = Y2(Y1)$ in Y3 with the keystrokes $\boxed{\text{VARS}}$ $\boxed{\blacktriangleright}$ 1 [Function] 2 [Y2] $\boxed{(}$ $\boxed{\text{VARS}}$ $\boxed{\blacktriangleright}$ 1 [Function] 1 [Y1] $\boxed{)}$ $\boxed{\text{ENTER}}$.

Next, enter your algebraic answer for the composite function in Y4. Be certain that you enclose the exponent in Y4 (the function in Y1) in parentheses!

(The composite function in the text is the one that appears to the right, but you should enter the function that you have found for the composite function.)

We now wish to check that the algebraic answer for the composite function is the same as the calculator's composite function by evaluating both functions at several different input values. You can do these evaluations on the home screen, but using the table is more convenient.

1.3.5 TURNING FUNCTIONS IN THE GRAPHING LIST OFF AND ON

Note that we are interested in the output values for only Y3 and Y4. However, the table will list values for all functions that are turned on. (A function is *turned on* when the equals sign in its location in the graphing list is darkened.) We now wish to turn off Y1 and Y2.

Press $\boxed{Y=}$ and, using one of the arrow keys, move the cursor until it covers the darkened = in Y1. Press $\boxed{\text{ENTER}}$. Then, press $\boxed{\blacktriangledown}$ until the cursor covers the darkened = in Y2. Press $\boxed{\text{ENTER}}$. Y1 and Y2 are now turned off.

A function is *turned off* when the equals sign in its location in the graphing list is not dark. To turn a function back on, simply reverse the above process to make the equal sign for the function dark. When you draw a graph, the TI-83 graphs of all functions that are turned on. You may at times wish to keep certain functions entered in the graphing list but not have them graph and not have their values shown in the table. Such is the case in this illustration. We now return to checking to see that Y3 and Y4 represent the same function.

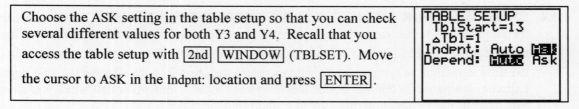

Choose the ASK setting in the table setup so that you can check several different values for both Y3 and Y4. Recall that you access the table setup with $\boxed{\text{2nd}}$ $\boxed{\text{WINDOW}}$ (TBLSET). Move the cursor to ASK in the Indpnt: location and press $\boxed{\text{ENTER}}$.

TI-83 Guide A-15

| Press [2nd] [GRAPH] (TABLE), type in the x-value(s) at which the function is to be evaluated, and press [ENTER] after each one. We see that because all these outputs are the same for each function, you can be fairly sure that your answer is correct. | 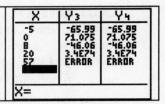 |

Why does ERROR appear in the table when $x = 57$? Look at the value when $x = 20$; it is very large! The computational limits of the calculator have been exceeded when $x = 57$. The TI-83 calls this an OVERFLOW error.

1.3.6 GRAPHING AN INVERSE FUNCTION The TI-83 can draw the graph of the inverse of a function. Using the calculator graph, you can check your algebraic answer. We illustrate this idea using the function in Example 4 of Section 1.3 of *Calculus Concepts*:

> The fares for a cab company are determined by the function $F(d) = 1.8d + 2.5$ dollars where d is the distance traveled in miles.

| Press [Y=], clear any previously entered functions, and enter the function F in Y1. Remember to use X as the input variable. (Refer to Sections 1.1.2 and 1.1.3 of this *Guide* for hints on how to set the WINDOW to graph this function.) | |

This problem does not state an interval of values for which the input variable is defined, so we begin by guessing one that makes sense in context. Because d is the distance traveled, we know that $d \geq 0$. We choose a maximum value of 50. (Your guess is as good as the one that is given below.)

Press [WINDOW] and enter appropriate values for the input and output. Draw the graph of the function for the view you choose.	WINDOW Xmin=0 Xmax=50 Xscl=5 Ymin=0 Ymax=95 Yscl=10 Xres=1	
Press [Y=] and enter in Y2 your answer to part *a* of Example 4 – the inverse function for F. Again, remember that the TI-83 requires that you use X as the input variable and that you use parentheses around any numerator and/or denominator that consists of more than one symbol in a fraction.	Plot1 Plot2 Plot3 \Y1■1.8X+2.5 \Y2■(X-2.5)/1.8 \Y3= \Y4= \Y5= \Y6= \Y7=	
Press [GRAPH]. If you do not have a good view of both functions, reset the window. Return to the home screen with [2nd] [MODE] (QUIT).		
Draw the TI-83 constructed inverse function by pressing [2nd] [PRGM] [DRAW] 8 [DrawInv] [VARS] [▶] 1 [Function] 1 [Y1] [ENTER]. If you do see a third graph, your inverse function formula is not correct.		

CAUTION: If you do not see a third graph, as is the case in the graph shown above, you may have computed the formula correctly. However, you might have an incorrect inverse function formula whose graph does not show in the current viewing window. We suggest that if you do not see a third graph that you turn off the graphing location where you have entered your inverse function formula – here Y2. (Remember that you turn off a function by having the equals sign in its location in the Y= list not darkened.) Then, on the home screen, press 2nd ENTER (ENTRY) ENTER to redraw the inverse function graph and visually check that it is the same graph as the one for Y2. Realize that these methods provide only a visual check on your answer and are not exact.

| The answers to parts *b* and *c* of Example 4 are found using the distance (inverse) function in Y2 and the fare (original) function in Y1, respectively. Don't forget to include units of measure with the answers. | Y2(33.1)   17
Y1(3)   7.9 |

1.3.7 COMPOSITION PROPERTY OF INVERSE FUNCTIONS This concept, involved in part *d* of Example 4, provides another check on your answer for the inverse function.

| Form the compositions of the inverse function and the original function according to the Composition Property of Inverse Functions and enter them in Y3 and Y4. Turn off the functions in Y1 and Y2. (Section 1.3.3 of this *Guide*.) | Plot1 Plot2 Plot3
\Y1=1.8X+2.5
\Y2=(X-2.5)/1.8
\Y3=Y2(Y1)
\Y4=Y1(Y2)
\Y5=
\Y6=
\Y7= |
| Have the TABLE set to ASK mode (See Section 1.1.6) and press 2nd GRAPH (TABLE). Enter several values for X to see if both composite functions return that value of X. If so, your answer for the inverse function is very likely correct. | X   Y3   Y4
-8   -8   -8
0   0   0
2.8 2.8 2.8
20.26 20.26 20.26
38   38   38
X= |

1.3.8 GRAPHING A PIECEWISE CONTINUOUS FUNCTION Piecewise continuous functions are used throughout the text. You will need to use your calculator to graph and evaluate outputs of piecewise continuous functions. Several methods can be used to draw the graph of a piecewise function. One of these is presented below using the function that appears in Example 6 of Section 1.3 in *Calculus Concepts*:

The population of West Virginia from 1985 through 1993 can be modeled by

$$P(t) = \begin{cases} -23.514t + 3903.667 \text{ thousand people} & \text{when } 85 \le t < 90 \\ 7.7t + 1098.7 \text{ thousand people} & \text{when } 90 \le t \le 93 \end{cases}$$

where *t* is the number of years since 1900.

Parentheses are used around each portion of the function and around an inequality statement that tells the calculator which side of the break point to graph each part of the function. We set the lower and upper endpoints of the input interval as Xmin and Xmax, respectively.

Clear any functions that are in the Y= list. Enter the first piece of P (using X as the input variable) in the Y1 location with the keystrokes [(] [(–)] 23.514 [X,T,θ,n] [+] 3903.667 [)] [(] [X,T,θ,n] [2nd] [MATH] [TEST] 5 [<] 90 [)] [ENTER].

Enter the second piece of the function P (using X as the input variable) in the Y2 location with [(] 7.7 [X,T,θ,n] [+] 1098.7 [)] [(] [X,T,θ,n] [2nd] [MATH] [TEST] 4 [≥] 90 [)].

Each piece of the function and its corresponding input at the break point must be inside parentheses.

Your calculator draws graphs by connecting function outputs wherever the function is defined. However, this piecewise function breaks at $x = 90$. The TI-83 will connect the two pieces of P unless you tell it not to do. Whenever you draw graphs of piecewise functions, you should set your calculator to Dot mode as described below so that it will not connect the different pieces of the function.

Have the cursor on the first line of the function and press [◄] until the blinking cursor is on the slanted line[2] to the *left* of Y1. Press [ENTER] six times. You should see the solid slanted line change shapes until it becomes a dotted slanted line (as shown to the right). Press [▼] and then press [ENTER] six times to put the function in Y2 in Dot mode. Press [►] several times to return to the right of the equals sign.

Note: The method described above places individual functions in Dot mode. The functions return to standard (connected) mode when the function locations are cleared. If you want to put all functions in Dot mode at the same time, press [MODE] [▼] [▼] [▼] [▼] [►] [Dot] [ENTER]. However, if you choose to set DOT mode in this manner, you must return to the MODE screen, select Connected, and press [ENTER] to take the TI-83 out of DOT mode.

Now, set the window. The function P is defined only when the input is between 85 and 93. So, we evaluate $P(85)$, $P(93)$, and $P(90)$ to help when setting the vertical view.

(Note that if you attempt to set the window using ZoomFit as described in Section 1.1.2 of this *Guide*, the picture is not very good and you need to manually reset the height of the window as described below.)

Press [WINDOW], set Xmin = 85, Xmax = 93, Ymin ≈ 1780, and Ymax ≈ 1905. Press [GRAPH].

---

[2] The different graph styles you can draw from this location are described in more detail on pages 3-9 through 3-10 in your *TI-83 Owner's Guidebook*.

Reset the window if you want a closer look at the function around the break point. The graph to the right was drawn using Xmin = 89, Xmax = 91, Ymin = 1780, and Ymax = 1810. (The breaks in the left piece of the graph occur because we are graphing in dot mode, not connected mode.)	
You can find function values by evaluating outputs on the home screen or using the table. Carefully look at the inequalities in the function P to determine whether Y1 or Y2 should be evaluated to obtain each particular output. (See Sections 1.1.5 and 1.1.6 for more detailed instructions on how to evaluate functions.)	Y₁(87) 1857.949 Y₂(92) 1807.1 Y₁(85) 1904.977 Y₂(90)▪

1.4 Limits: Describing Function Behavior

The TI-83 table is an essential tool when you are estimating limits numerically. Even though rounded values are shown in the table due to space limitations, the TI-83 displays at the bottom of the screen many more decimal places for a particular output when you highlight that output.

1.4.1 ESTIMATING LIMITS NUMERICALLY Whenever you use the TI-83 to estimate limits numerically, you will find it easiest to set the TABLE to ASK mode. We illustrate using the function u that appears in Example 2 of Section 1.4 in *Calculus Concepts*:

Press $\boxed{Y=}$ and use $\boxed{CLEAR}$ to delete all previously-entered functions. Enter $u(x) = \dfrac{3x^2 + 3x}{9x^2 + 11x + 2}$. Be certain to enclose the numerator and denominator of the fraction in parentheses.	Plot1 Plot2 Plot3 \Y₁■(3X²+3X)/(9X²+11X+2) \Y₂= \Y₃= \Y₄= \Y₅= \Y₆=
Press $\boxed{2nd}$ $\boxed{WINDOW}$ (TBLSET). Choose Ask in the Indpnt: location by placing the cursor over Ask and pressing $\boxed{ENTER}$. (Remember that the other settings on this screen do not matter if you are using Ask.)	TABLE SETUP TblStart=18 △Tbl=1 Indpnt: Auto **Ask** Depend: **Auto** Ask
Press $\boxed{2nd}$ $\boxed{GRAPH}$ (TABLE). Delete any values that are there with $\boxed{DEL}$. To numerically estimate $\lim\limits_{x \to -1^-} u(x)$, enter values to the *left* of, and becoming *closer and closer* to, -1. (*Note*: The values you enter do not have to be those shown in the text or these shown to the right.)	X | Y₁ -1.5 | .3913 -1.05 | .42282 -1.005 | .42796 **-1.0005** | .42851 -1 | .42857 -1 | .42857 X=-1.0005

CAUTION: Your instructor will very likely have you write the table you construct on paper. The TI-83 rounds off the table values it displays because of space limitations. Be certain to highlight the values and look on the bottom of the screen to see what these values actually are. The inputs that were entered in the table above are -1.5, -1.05, -1.005, -1.0005, -1.00005, and -1.000005. The outputs also have been rounded off.

ROUNDING OFF: Recall that *rounded off* means that if one digit past the digit of interest if less than 5, other digits past the digit of interest are dropped. If one digit past the one of interest is 5 or over, the digit of interest is increased by 1 and the remaining digits are dropped. (For instance, the highlighted number in the last screen shown above, -1.0005, has been rounded off by the TI-83 to 3 decimal places. Because the 5 in the fourth decimal place location is in the "5 or over" category, the 0 before this 5 is increased by 1 with the digits past that point dropped to give -1.001.)

RULE OF THUMB FOR DETERMINING LIMITS FROM TABLES: Suppose that we want $\lim_{x \to -1^-} u(x)$ accurate to 3 decimal places. Watch the table until you see that the output is the same value to one more decimal place (here, to 4 decimal places) for three consecutive outputs. Then, round that common value off to the requested 3 places for the desired limit. Your instructor may establish a different rule from this one, so be sure to ask.

Using this Rule of Thumb and the results that are shown on the last calculator screen, we estimate that $\lim_{x \to -1^-} u(x) = 0.429$. We now need to estimate the limit from the right of -1.

Delete the values currently in the table with DEL. To numerically estimate $\lim_{x \to -1^+} u(x)$, enter values to the *right* of, and becoming *closer and closer* to, -1. (*Note*: Again, the values that you enter do not have to be those shown in the text or these shown to the right.)	X: $-.98$, $-.998$, $-.9998$, -1, -1; Y1: $.43109$, $.42882$, $.4286$, $.42857$, $.42857$. X= $-.999998$
Because the output $0.4285\ldots$ appears three times in a row, we estimate that $\lim_{x \to -1^+} u(x) = 0.429$. Then, because $\lim_{x \to -1^-} u(x) = \lim_{x \to -1^+} u(x) = 0.429$, we estimate that $\lim_{x \to -1} u(x) = 0.429$.	X: $-.98$, $-.998$, $-.9998$, -1, -1; Y1: $.43109$, $.42882$, $.4286$, $.42857$, $.42857$. Y1 = $.428595924666$

We now illustrate finding the limit in part *b* of Example 2 in Section 1.4 of *Calculus Concepts*:

Delete the values currently in the table with DEL. To numerically estimate $\lim_{x \to -2/9^-} u(x)$, enter values to the *left* of, and becoming *closer and closer* to, $-2/9 = -0.222222\ldots$. Because the output values appear to become larger and larger, we estimate that $\lim_{x \to -2/9^-} u(x) \to \infty$.	X: $-.23$, $-.223$, $-.2223$, $-.2222$, $-.22222$; Y1: 9.8571, 95.571, 952.71, 9524.1, 95238. X = $-.222223$
Delete the values currently in the table. To numerically estimate $\lim_{x \to -2/9^+} u(x)$, enter values to the *right* of, and becoming *closer and closer* to, $-2/9$. Because the output values appear to become larger and larger, we estimate $\lim_{x \to -2/9^-} u(x) \to -\infty$. We conclude that $\lim_{x \to -2/9} u(x)$ does not exist.	X: $-.21$, $-.218$, $-.2218$, $-.2222$, $-.22222$; Y1: -5.727, -17.21, -175.1, -1754, -17544. X = $-.222218$

1.4.2 CONFIRMING LIMITS GRAPHICALLY – ZOOMING IN AND OUT
A graph can be used to confirm a limit that you estimated numerically. You also can *zoom in* or *zoom out* on the graph to obtain a better view of the limit you are estimating. We again illustrate using the function *u* that appears in Example 2 of Section 1.4 in *Calculus Concepts*.

Have the function $u(x) = \dfrac{3x^2 + 3x}{9x^2 + 11x + 2}$ entered in the Y1 location of the Y= list. A graph drawn with ZOOM 4 [ZDecimal] or ZOOM 6 [ZStandard] is not very helpful. (Refer to Sections 1.1.2 and 1.1.3 of this *Guide* for hints on how to set the WINDOW to graph this function.)	
To confirm that $\lim\limits_{x \to {}^-1} u(x) = 0.429$, we are only interested in values of u that are near -1. So, choose values very near to -1 for the x-view and evaluate the function at those x-values to help determine the y-view. We manually set the window to values such as those shown to the right and draw the graph.	
If you look closely, you can actually see the "hole" in the graph at $x = -1$. Press TRACE in this window, the TI-83 confirms that u is not defined at $x = -1$. After pressing TRACE, press ▶ and observe the y-value. Press ◀ several times to go to the left of $x = -1$ and confirm from the y-values that the limit is approximately what we had determined numerically.	

Note that we confirmed that the limit exists by seeing that the two parts of the graph (to the left and right of $x = -1$) move toward the same point. Tracing around $x = -1$ provides a check on the numerical value of the limit.

The previous instructions show how to *zoom in* by manually setting the window. You can also zoom in with the ZOOM menu of the calculator. We next describe this method.

Return to the graphing screen set with ZOOM 4 [ZDecimal] (or ZOOM 6 [ZStandard] or any screen where you can see the portion of the graph around $x = -1$.) Press ZOOM 2 [Zoom In] and use ◀ and ▲ to move the blinking cursor until you are near the point on the graph where $x = -1$. Press ENTER.	
If you look closely, you can see the hole in the graph at $x = -1$. If this view is not magnified enough to see what is happening around $x = -1$, press ENTER to zoom in again. To check your numerical estimate of the limit, press TRACE, use the arrow keys to move to either side of $x = -1$, and observe the y-values.	

We now consider the limit in part *b* of Example 2 of Section 1.4. We want to confirm with a graph what was found numerically; that is, we wish to confirm that $\lim\limits_{x \to {}^-2/9} u(x)$ does not exist. However, first we should eliminate the "extra" vertical line that appears on the above graph at

$x = -2/9$. This line appears because we are graphing in Connected mode (which tells the TI-83 to connect points on the graph). Note that you may or may not have the line on your graph.

Have Y1 set to draw in Dot mode. (See page A-17 for the keystrokes needed to put the TI-83 in Dot mode.) If you use Option 3 in the ZOOM menu (Zoom Out), the graph is lost. We therefore manually set the window.	Plot1 Plot2 Plot3 \Y1∎(3X²+3X)/(9X²+11X+2) \Y2= \Y3= \Y4= \Y5= \Y6=
We want to zoom out on the graph to confirm that the limit of $u(x)$ as x approaches $-2/9$ does not exist. To do this, we set a very small x-view and a larger y-view. (Note that these values are arbitrary – any small x-view that includes $-2/9$ and any y-view in which the graph can be seen clearly will do.)	WINDOW Xmin=-.23 Xmax=-.21 Xscl=1 Ymin=-50 Ymax=50 Yscl=10 Xres=1
You can TRACE the graph to confirm that where it "splits" is near $x = -2/9 \approx -0.222222\ldots$. It appears that as x approaches $-2/9$ from the left that the outputs become larger and larger and that as x approaches $-2/9$ from the right that the outputs become smaller and smaller.	

1.4.3 INVESTIGATING END BEHAVIOR NUMERICALLY
Investigating end behavior using the TI-83 table is very similar to numerically estimating the limit at a point. We illustrate this using the function that appears in Example 4 of Section 1.4 in *Calculus Concepts*.

Have $u(x) = \dfrac{3x^2 + x}{3x + 11x^2 + 2}$ in the Y1 location of the Y= list. Be certain that you remember to enclose both the numerator and denominator of the fraction in parentheses.	Plot1 Plot2 Plot3 \Y1∎(3X²+X)/(3X+11X²+2) \Y2= \Y3= \Y4= \Y5= \Y6=
Have TBLSET set to Ask and press [2nd] [GRAPH] (TABLE). Delete any values that are in the table with [DEL]. In order to numerically estimate $\lim_{x \to \infty} u(x)$, enter values of x that get larger and larger. (*Note*: The values that you enter do not have to be those shown in the text or these shown to the right.)	X | Y1 200 | .27281 800 | .27275 1500 | .27274 5000 | .27273 10000 | Y1=.272728925079
We assume that we want the limit accurate to 3 decimal places. According to our *Rule of Thumb for Determining Limits from Tables* (page A-19), once we see the same value to 4 decimal positions 3 times in a row, we can estimate the limit by rounding off the answer to 3 decimal places.	$\lim_{x \to \infty} u(x) \approx 0.273$ Note that the last entry in the table ($x = 10{,}000$) was not necessary.
Part b of Example 4 asks for $\lim_{x \to -\infty} u(x)$. Delete the values that are currently in the table with [DEL]. Then, enter values of x that get smaller and smaller. We estimate $\lim_{x \to -\infty} u(x) \approx 0.273$.	X | Y1 -300 | .27267 -900 | .27271 -5000 | .27272 -10000 | Y1=.272725619294

CAUTION: It is not the final value, but a sequence of several values, that is important when determining limits. If you enter a very large or very small value, you may exceed the limits of the TI-83's capability and obtain an incorrect number. Always look at the sequence of values obtained to make sure that all values found make sense.

1.4.4 INVESTIGATING END BEHAVIOR GRAPHICALLY As was the case with limits at a point, a graph of the function can be used to confirm a numerically-estimated limit. We again illustrate with the function that appears in Example 4 of Section 1.4 in *Calculus Concepts*.

Have $u(x) = \dfrac{3x^2 + x}{3x + 11x^2 + 2}$ in the Y1 location of the Y= list. (Be certain that you remember to enclose both the numerator and denominator of the fraction in parentheses.) A graph drawn with ZOOM 4 [ZDecimal] is a starting point.	
We estimated the limit as x gets very large or very small to be 0.273. Now, $u(0) = 0$, and it does appears from the graph that u is never negative. Set a WINDOW with values such as Xmin = −10, Xmax = 10, Ymin = 0, and Ymax = 0.35. Press GRAPH.	
To examine the limit as x gets larger and larger, change the WINDOW so that Xmax = 100, view the graph with GRAPH, change the WINDOW so that Xmax = 1000, view the graph with GRAPH, and so forth. The graph to the right was drawn with Xmax = 10,000. Press TRACE on each graph screen.	
Repeat the process as x gets smaller and smaller, but change Xmin rather than Xmax after drawing each graph. The graph to the right was drawn with Xmin = −9000, Xmax = 10, Ymin = 0, and Ymax = 0.35. Again, press TRACE on each graph screen to view some of the outputs and confirm the numerical estimates.	

1.5 Linear Functions and Models

This portion of the *Guide* gives instructions for entering real-world data into the calculator and finding familiar function curves to fit that data. You will use the beginning material in this section throughout all the chapters in *Calculus Concepts*.

CAUTION: Be very careful when you enter data in your calculator because your model and all of your results depend on the values that you enter! Always check your entries.

1.5.1 ENTERING DATA We illustrate data entry using the values in Table 1.46 in Section 1.5 of *Calculus Concepts*:

Year	1996	1997	1998	1999	2000	2001
Tax (in dollars)	2532	3073	3614	4155	4696	5237

Press STAT 1 [EDIT] to access the six lists that hold data. You see only the first three lists, (L1, L2, and L3) but you can access the other three lists (L4, L5, and L6) by having the cursor

TI-83 Guide

on the list name and pressing ▶ several times. If you do not see L1, L2, and so forth, return to the statistical setup instructions on page A-1 of this *Guide*.

In this text, we usually use list L1 for the input data and list L2 for the output data. If there are any data values already in your lists, see Section 1.5.3 of this *Guide* and first delete any "old" data. To enter data in the lists, do the following:

> Position the cursor in the first location in list L1. Enter the input data into list L1 by typing the years from top to bottom in the L1 column, pressing [ENTER] after each entry.
>
> After typing the sixth input value, 2001, use ▶ to cause the cursor go to the top of list L2. Enter the output data into list L2 by typing the entries from top to bottom in the L2 column, pressing [ENTER] after each tax value.

1.5.2 EDITING DATA If you incorrectly type a data value, use the cursor keys (*i.e.*, the arrow keys ▶, ◀, ▲, and/or ▼) to darken the value you wish to correct and then type the correct value. Press [ENTER].

- To *insert* a data value, put the cursor over the value that will be directly below the one you will insert, and press [2nd] [DEL] (INS). The values in the list below the insertion point move down one location and a 0 is filled in at the insertion point. Type the data value to be inserted over the 0 and press [ENTER]. The 0 is replaced with the new value.

- To *delete* a single data value, move the cursor over the value you wish to delete, and press [DEL]. The values in the list below the deleted value move up one location.

1.5.3 DELETING OLD DATA Whenever you enter new data in your calculator, you should first delete any previously-entered data. There are several ways to do this, and the most convenient method is illustrated below.

> Access the data lists with [STAT] 1 [EDIT]. (You probably have different values in your lists if you are deleting "old" data.) Use ▲ to move the cursor over the name L1. Press [CLEAR] [ENTER].
>
> Use ▶ ▲ to move the cursor over the name L2. Press [CLEAR] [ENTER]. Repeat this procedure to clear old data from any of the other lists you will use.

1.5.4 FINDING FIRST DIFFERENCES When the input values are evenly spaced, you can use program DIFF to compute first differences in the output values. Program DIFF is given in the *TI-83 Program Appendix* at the *Calculus Concepts Website*. Consult the Programs category in *TroubleShooting the TI-83* in this *Guide* if you have questions about obtaining the programs.

Have the data given in Table 1.46 in Section 1.5 of *Calculus Concepts* entered in your calculator. (See Section 1.5.1 of this *Guide*.) Exit the list menu with [2nd] [MODE] (QUIT).	![L1: 1996-2001, L2: 2532, 3073, 3614, 4155, 4696, 5237]
To run the program, press [PRGM] followed by the number that is to the left of the DIFF program location, and press [ENTER]. The message on the right appears on your screen. (*Note:* We use the information in lists L4 and L5 in the next chapter.)	HAVE X IN L1 HAVE Y IN L2–SEE 1ST DIFF IN L3, 2ND DIFF IN L4, PERCENT CHANGE IN L5 　　　　　Done
Press [STAT] 1 [EDIT] to view the first differences in list L3. The first differences in L3 are constant at 541, so a linear function gives a perfect fit to these tax data.	L3: 541, 541, 541, 541, 541 L3 = {541,541,541...

You will find program DIFF very convenient in the next chapter because of other options it has. However, if you do not want to use program DIFF at this time, you can use a built-in capability of your TI-83 to compute the first differences of any list. Be certain that you have the output data entered in list L2 in your calculator.

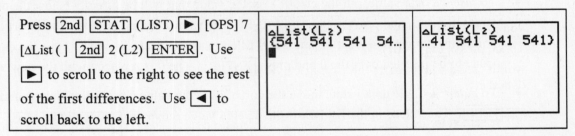

Press [2nd] [STAT] (LIST) [▶] [OPS] 7 [ΔList ( ] [2nd] 2 (L2) [ENTER]. Use [▶] to scroll to the right to see the rest of the first differences. Use [◀] to scroll back to the left.

Note: Program DIFF **should not** be used for data with input values (L1) that are *not* evenly spaced. First differences give no information about a possible linear fit to data with inputs that are not the same distance apart. If you try to use program DIFF with input data that are not evenly spaced, the message INPUT VALUES NOT EVENLY SPACED appears and the program stops. The ΔList option discussed above gives first differences even if the input values are not evenly spaced. However, these first differences still have no interpretation when finding a linear function to fit data if the input values are not evenly spaced.

1.5.5 SCATTER PLOT SETUP The first time that you draw a graph of data, you need to set the TI-83 to draw the type of graph you want to see. Once you do this, you never need to do this set up again (unless for some reason the settings are changed). If you always put input data in list L1 and output data in list L2, you can turn the scatter plots off and on from the Y= screen rather than the STAT PLOTS screen after you perform this initial setup.

Press [2nd] [Y=] (STAT PLOT) to display the STAT PLOTS screen. (Your screen may not look exactly like this one.)	STAT PLOTS 1:Plot1...Off 　L1　L2 2:Plot2...Off 　L3　L4 3:Plot3...Off 　L5　L6 4↓PlotsOff

On the STAT PLOTS screen, press ENTER to display the Plot1 screen, press ENTER to turn Plot1 on, and choose the options shown on the right. (You can choose any of the 3 marks at the bottom of the screen.)	Plot1 Plot2 Plot3 On **Off** Type: ▦ ⊵ ⊞ ⊞ ⊞ ⊵ Xlist:L1 Ylist:L2 Mark: □ + ·
Press Y= and notice that Plot1 at the top of the screen is dark. This tells you that Plot1 is turned on and ready to be graphed. Press 2nd MODE (QUIT) to return to the home screen.	**Plot1** Plot2 Plot3 \Y1=■ \Y2= \Y3= \Y4= \Y5= \Y6= \Y7=

- A scatter plot is *turned on* when its name on the y(x)= screen is darkened. From now on, you can turn the scatter plot off and on by having Plot1 either not dark or dark. To turn Plot1 off, use ▲ to move the cursor to the Plot1 location, and press ENTER. Reverse the process to turn Plot1 back on.

- TI-83 lists can be named and stored in the calculator's memory for later recall and use. Refer to Section 1.5.13 and 1.5.14 of this *Guide* for instructions on storing data lists and later recalling them for use.

CAUTION: Anytime and for whatever reason you enter the name of a numbered list (for instance: L1, L2, and so forth), you should use the TI-83 symbol for the name, **not** a name that you type with the alphabetic and numeric keys. For example, to enter the list name L1 on the TI-83, press 2nd 1 (L1). If you instead entered L1 with the keystrokes ALPHA) (L) 1, you would create a new list called L1 that the calculator would not recognize as the list containing data that you enter in it while in the STAT [EDIT] mode.

The screen to the right contains three entries marked L1. The first is the statistical list L1 that contains input data. See what is in your L1 list by pressing 2nd 1 (L1) ENTER.
 	
The second entry is obtained by pressing ALPHA) (L) 1 ENTER. Whatever constant is stored in memory location L multiplied by 1 is what you see printed on the screen.
 	
If you stored numbers in a list that you created and named L1 using ALPHA) (L) 1 ENTER, you will see the third entry shown on the screen to the right when you press 2nd STAT (LIST) followed by the number corresponding to the location of L1 and ENTER.
	L1
	{1996 1997 1998…
	L1
	                      25
	ᴌL1
	                   {0 2}
	
	Your list L1 may not contain the same values as the one shown above. You may not have the third entry shown above in your calculator. If you do have it, it will contain different values.

If you have the third entry that is shown on the screen above (ᴌL1) or if you see any lists that begin with the letter L followed by a large number (not a small number which denotes a built-in list) when you press 2nd STAT (LIST), refer to Section 1.5.15 of this *Guide*. These lists need to be deleted or you will probably be using incorrect data for everything!

1.5.6 DRAWING A SCATTER PLOT OF DATA

Any functions that are turned on in the Y= list will graph when you plot data. Therefore, you should clear or turn them off before you draw a scatter plot. We illustrate how to graph data using the modified tax data that follows Example 2 in Section 1.5 of *Calculus Concepts*.

Year	1996	1997	1998	1999	2000	2001
Tax (in dollars)	2541	3081	3615	4157	4703	5242

Access the Y= graphing list. If any entered function is no longer needed, clear it by pressing CLEAR. If you want the function(s) to remain but not graph when you draw the scatter plot, refer to Section 1.3.4 of this *Guide* for instructions on how to turn function(s) off. Also be sure that Plot 1 at the top left of the Y= screen is darkened.

Press STAT 1 [EDIT]. Using the table given above, enter the year data in L1 and the modified tax data in L2 according to the instructions given in Section 1.5.1 of this *Guide*. (You can either leave values in the other lists or clear them.)

Press ZOOM 9 [ZoomStat] to have the calculator set an auto-scaled view of the data and draw the scatter plot.

Note that ZoomStat does not reset the *x*- and *y*-axis tick marks. You should do this manually with the WINDOW settings if you want different spacing between the tick marks.)

Recall that if the data are perfectly linear (that is, every data point falls on the graph of a line), the first differences in the output values are constant. The first differences for the original tax data were constant at $541, so a linear function fit the data perfectly. What information is given by the first differences for these modified tax data?

Run program DIFF by pressing PRGM followed by the number that is to the left of the DIFF program location, and press ENTER. Press STAT 1 [EDIT] to view the first differences in list L3.

These first differences are close to being constant. This information, together with the linear pattern shown by the scatter plot, are a good indication that a linear function is likely to give a good fit to the data.

1.5.7 FINDING A LINEAR FUNCTION TO MODEL DATA

You will often have your TI-83 find the linear function that best fits a set of data. Your calculator can find two different, but equivalent, forms of a linear equation: $y = ax + b$ or $y = a + bx$. For convenience, we always choose the equation $y = ax + b$.

Press 2nd MODE (QUIT) to return to the home screen. Then, press STAT ▶ [CALC] 4 [LinReg(ax+b)] to access the TI-83's "fit a linear function to the data" instruction.

TI-83 Guide

Copy the fit instruction to the home screen with [ENTER]. Also paste the equation into the Y1 location of the Y= list by pressing [VARS] [▶] [Y-VARS] 1 [Function] 1 [Y1]. Press [ENTER].	`LinReg(ax+b) Y₁`	`LinReg`
		`y=ax+b`
		`a=540.3714286`
		`b=-1076042.467`

CAUTION: The best-fit function found by the calculator is also called a *regression function*. The coefficients of the regression function ***never*** *should be rounded*! This is not a problem because the calculator pastes the entire equation it finds into the Y= list at the same time the function is found if you follow the instructions given above.

The linear equation of best fit for the modified tax data that was entered into lists L1 and L2 in Section 1.5.6 and displayed on the home screen has also been copied into Y1. Press [Y=] to verify this.	`Plot1 Plot2 Plot3`
	`\Y₁■540.37142857`
	`145X+-1076042.46`
	`66667`
	`\Y₂=`
	`\Y₃=`
	`\Y₄=`
	`\Y₅=`

NOTE: When finding the equation of best fit, the TI-83 will by default use L1 as the input data list and L2 as the output data list unless you tell it do otherwise. It is possible to use lists other than L1 and L2 for the input and output data. However, if you do so, you must set one of the STAT PLOT locations to draw the scatter plot for those other lists (as described in Section 1.5.5). To paste the function into a location other than Y1, just change the number 1 following y in the instructions given above to the number corresponding to the location you want. For example, these keystrokes find a linear function for input data in list L3 and output data in list L4, and paste the equation into graphing location Y2: [STAT] [▶] [CALC] 4 [LinReg(ax+b)] [2nd] 3 (L3) [,] [2nd] 4 (L4) [,] [VARS] [▶] [Y-VARS] 1 [Function] 2 [Y2] [ENTER].

CAUTION: If you see a number that is labeled r on the screen that shows the linear function you fit to the data, tell the TI-83 to not show this value again by pressing (on the home screen) [2nd] 0 (CATALOG) [x^{-1}] and press [▼] until you the cursor arrow is next to DiagnosticOff. Press [ENTER] [ENTER]. The r that is shown is called the *correlation coefficient*. This and a quantity called r^2, the *coefficient of determination*, are numbers that you will learn about in a statistics course. It is not appropriate[3] to use these values in a calculus course.

Graphing the Line of Best Fit: After finding a best-fit equation, you should always draw the graph of the function on a scatter plot to verify that the function provides a good fit to the data.

Press [GRAPH] to overdraw the function you pasted in the Y= list on the scatter plot of the data. (As we suspected from looking at the scatter plot and the first differences, this function provides a very good fit to the data.)	

---

[3] Unfortunately, there is no single number that can be used to tell whether one function better fits data than another. The correlation coefficient only compares linear fits and should not be used to compare the fits of different types of functions. For the statistical reasoning behind this statement, read the references in footnote 6 on page A-32.

1.5.8 USING A MODEL FOR PREDICTIONS You can use one of the methods described in Section 1.1.5 or Section 1.1.6 of this *Guide* to evaluate the linear function at the indicated input value.

Predict the tax owed in 2002 where the tax is found using the linear model computed in the last section of this *Guide:* Tax = $540.371t - 1{,}076{,}042.467$ dollars where t is the year. Remember that you should always use the full model, *i.e.*, the function you pasted in Y1, not the rounded equation given above, for all computations.	Y1(2002) 5781.133333
The 2002 tax, predicted by evaluating an output on the home screen, is about \$5781. We now predict the tax in 2003 using the TI-83 table. As seen to the right, the predicted tax is approximately \$6322.	X Y1 2003 ▮▮▮▮▮ Y1=6321.5047619

1.5.9 COPYING A GRAPH TO PAPER Your instructor may ask you to copy what is on your graphics screen to paper. If so, use the following ideas to more accurately perform this task.

After using a ruler to place and label a scale (*i.e.*, tick marks) on your paper, use the trace values (as shown below) or the data to graph a scatter plot on your paper.

Press GRAPH to return the modified tax data graph found in Section 1.5.7 to the screen. Press TRACE and ▶. The symbols P1:L1,L2 in the upper left-hand corner of the screen indicate that you are tracing the scatter plot of the data.	P1:L1,L2 X=1997 Y=3081
Press ▼ to move the trace cursor to the linear function graph. The equation at the top of the screen is that of the function that you are tracing (in this case, Y1). Use ▶ and/or ◀ to locate values that are as "nice" as possible and mark those points on your paper. Use a ruler to connect the points and draw the line.	Y1=540.37142857145X+ -10 X=1998.883 Y=4096.7841

- If you are copying the graph of a continuous curve rather than a straight line, you need to trace as many points as necessary to see the shape of the curve while marking the points on your paper. Connect the points with a smooth curve.

1.5.10 ALIGNING DATA We return to the modified tax data entered in Section 1.5.6. If you want L1 to contain the number of years after a certain year instead of the actual year, you need to *align* the input data. In this illustration, we shift all of the data points to three different positions to the left of where the original values are located.

Press STAT 1 [EDIT] to access the data lists. To copy the contents of one list to another list; for example, to copy the contents of L1 to L3, use ▲ and ▶ to move the cursor so that L3 is highlighted. Press 2nd 1 (L1) ENTER.

Note: This first step shown above is not necessary, but it will save you the time it takes to re-enter the input data if you make a mistake. Also, it is not necessary to first clear L3. However, if you want to do so, have the symbols L3 highlighted and press CLEAR ENTER.

To align the input data as the number of years past 1996, first press the arrow keys (◄ and ▲) so that L1 is highlighted. Tell the TI-83 to subtract 1996 from each number in L1 with 2nd 1 (L1) − 1996.	L1 = L1 − 1996 with L1 values 1996–2001, L2 values 2541, 3081, 3615, 4157, 4703, 5242, L3 values 1996–2001
Press ENTER. Instead of an actual year, the input now represents the number of years since 1996. Return to the home screen with 2nd MODE (QUIT).	L1 values 0–5, L2 values 2541, 3081, 3615, 4157, 4703, 5242, L3 values 1996–2001; L1(1)=0
Find the linear function by pressing 2nd ENTER (ENTRY) as many times as needed until you see the linear regression instruction. To enter this function in a different location, say Y2, press ◄ and then press VARS ► [Y-VARS] 1 [Function] 2 [Y2].	LinReg(ax+b) Y2
Press ENTER and then press Y= to see the function pasted in the Y2 location. *Note:* If you want the aligned function to be in Y1, do not replace Y1 with Y2 before pressing ENTER to find the equation.	\Y1=540.37142857145X+ −1076042.46 66667 \Y2=540.37142857145X+2538.904761 9047 \Y3=
To graph this equation on a scatter plot of the aligned data, first turn off the function in Y1 (see page A-14 of this *Guide*) and then return to the home screen with 2nd MODE (QUIT). Press ZOOM 9 [ZoomStat].	(scatter plot with linear fit)

If you now want to find the linear function that best fits the modified tax data using the input data aligned another way, say as the number of years after 1900, first return to the data lists with STAT 1 [EDIT] and highlight L1.

Add 96 to each number currently in L1 with 2nd 1 (L1) + 96. Press ENTER. Instead of an actual year, the input now represents the number of years since 1900.		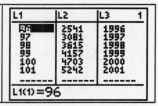

- There are many ways that you can enter the aligned input into L1. Another method that you can use is to start over from the beginning. Replace L1 with the contents of L3 by highlighting L1 and pressing 2nd 3 (L3) ENTER. Once again highlight the symbols L1 and subtract 1900 from each number in L1 with 2nd 1 (L1) − 1900.

On the home screen, find the linear function for the aligned data by pressing 2nd ENTER (ENTRY) until you see the linear regression instruction. To enter this new equation in a different location, say Y3, press ◄ and then press VARS ► [Y-VARS] 1 [Function] 3 [Y3]. Press ENTER and then press Y= to see the function pasted in the Y3 location.

- To graph this equation on a scatter plot of the aligned data, first turn off the other functions and then return to the home screen with 2nd MODE (QUIT). Press ZOOM 9 [ZoomStat].

- Remember, if you have aligned the data, the input value at which you evaluate the function may not be the value given in the question you are asked. You can use any of the equations to evaluate function values.

Using the function in Y1 (the input is the year), in Y2 (the input is the number of years after 1996), or in Y3 (the input is the number of years after 1900), we predict that the tax owed in 2003 is approximately $6322.

1.5.11 NAMING AND STORING DATA

You can name data (either input, output, or both) and store it in the calculator memory for later recall. You may or may not want to use this feature; it would be helpful if you plan to use large data sets several times. To illustrate the procedure, let's name the output data.

Press 2nd MODE (QUIT) to return to the home screen. You can view any list from the STAT EDIT mode (where the data is originally entered) or from the home screen by typing the name of the list and pressing ENTER. View the modified tax data in L2 with 2nd 2 (L2) ENTER.

Pressing ► allows you to scroll through the list to see the portion that is not displayed.
To store this data with the name TAX, press 2nd 2 (L2) STO► 2nd ALPHA TAX ENTER.

CAUTION: Do not store data to a name that is routinely used by the TI-83. Such names are L1, L2, ..., L6, MATH, LOG, MODE, A, B, and so forth. Note that if you use a single letter as a name, this might cause one or more of the programs to not execute properly.

1.5.12 RECALLING STORED DATA

The data you have stored remains in the memory of the TI-83 until you delete it using the instructions given in Section 1.5.15 of this *Guide*. When you wish to use the stored data, recall it to one of the lists L1, L2, ..., L5. We illustrate with the list named TAX, which we store in L2. Press 2nd STAT (LIST) and under NAMES, find TAX.

TI-83 Guide   A-31

| Press the number* corresponding to the location of the list. Press STO▶ 2nd 2 (L2) ENTER. List L2 now contains the TAX data. | NAMES OPS MATH
1:RESID
2:TAX | LTAX→L₂
{2541 3081 3615... |

*An equivalent action is to use ▼ to highlight the number and press ENTER.

1.5.13 DELETING DATA You do not need to delete any data lists unless your memory is getting low or you just want to do it. One exception to this is that you should delete any list that begins with the letter *L* and ends in a single *large* (*i.e.*, a large-size type) number. If you see a list beginning with *L* (such as L1 on the first screen below), you should delete it because the TI-83 may use the data in this list instead of the data you enter in the first list in the STAT EDIT mode. For illustration purposes, we delete lists L1 and TAX.

Press 2nd STAT (LIST). Then press 2nd + (MEM) 2 [Delete] ENTER.	NAMES OPS MATH 1:L1 2:RESID 3:TAX	MEMORY 1:Check RAM... 2:Delete... 3:Clear Entries 4:ClrAllLists 5:Reset...
Next, press 4: [List] and use ▼ to move the cursor opposite L1 (or TAX). DO NOT delete the TI-83's list L1 (*L* followed by a small, not large, 1). If you did not see a list L1 when 2nd STAT (LIST) was pressed, you only have the list TAX to delete.		DELETE:List L₃ 63 L₄ 45 L₅ 54 L₆ 54 ▶L1 28 RESID 67 TAX 65
Press ENTER. To delete another list, use ▼ to move the cursor opposite that list name and press ENTER. Exit this screen with 2nd MODE (QUIT) when finished.		DELETE:List L₃ 63 L₄ 45 L₅ 54 L₆ 54 ▶RESID 67

1.5.14 WHAT IS "BEST FIT"? It is important to understand the method of least squares and the conditions necessary for its use if you intend to find the equation that best-fits a set of data. You can explore[4] the process of finding the line of best fit with program LSLINE. (Program LSLINE is given in the *TI-83 Program Appendix*.) For your investigations of the least-squares process with this program, it is better to use data that is not perfectly linear and data for which you do *not* know the best-fitting line.

Before using program LSLINE, clear all functions from the Y= list, turn on Plot1 by darkening the name Plot1 on the Y= screen, and enter your data in lists L1 and L2. (If Plot 1 is not turned on, the program will not execute properly.) Next, draw a scatter plot with ZOOM 9 [ZoomStat]. Press WINDOW and reset Xscl and Yscl so that you can use the tick marks to help identify points when you are asked to give the equation of a line to fit the data. Press GRAPH, view the scatter plot, and then return to the home screen.

---

[4] This program works well with approximately 5 data points. Interesting data to use in this illustration are the height and weight, the arm span length and the distance from the floor to the navel, or the age of the oldest child and the number of years the children's parents have been married for 5 randomly selected persons.

Copyright © Houghton Mifflin Company. All rights reserved.

| To activate program LSLINE, press PRGM followed by the number corresponding to the location of the program, and press ENTER. The program first displays the scatter plot you constructed and pauses for you to view the screen. | |

NOTE: While the program is calculating, there is a small vertical line in the upper-right hand corner of the graphics screen that is dashed and "wiggly". This program pauses several times during execution for you to view the screen. Whenever the program pauses, the small line is "still" and you should press ENTER to resume execution after you have looked at the screen.

| The program next asks you to find the *y*-intercept and slope of *some* line you estimate will go "through" the data. (You should not expect to guess the best-fit line on your first try!) Use the tick marks to estimate rise divided by run and note a possible *y*-intercept. Enter your guess for the slope and *y*-intercept. | 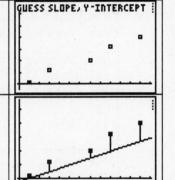 |
| After pressing ENTER again, your line is drawn and the errors are shown as vertical line segments on the graph. (You may have to wait a moment to see the vertical line segments before again pressing ENTER.) | |

Next the sum of squared errors, SSE, is displayed for your line. Decide whether you want to move the *y*-intercept of the line or change its slope to improve the fit to the data.

| Press ENTER and enter 1 to choose the TRY AGAIN? option. After entering another guess for the *y*-intercept and/or slope, the process of viewing your line, the errors, and display of SSE is repeated. | | 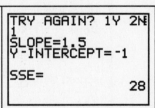 |

If the new value of SSE is smaller than the SSE for your first guess, you have improved the fit.

| When you feel an SSE value close to the minimum value is found, enter 2 at the TRY AGAIN? prompt. The program then overdraws the line of best fit on the graph and shows the errors for the line of best fit. | |  |

The program ends by displaying the coefficients *a* and *b* of the best-fit line $y = ax + b$ as well as the minimum SSE. Press ENTER to end the program. Use program LSLINE to explore[5] the method of least squares[6] that the TI-83 uses to find the line of best fit.

---

[5] Program LSLINE is for illustration purposes only. Actually finding the line of best fit for a set of data should be done according to the instructions in Section 1.5.7 of this *Guide*.

[6] Two articles that further explain "best-fit" are H. Skala, "Will the Real Best Fit Curve Please Stand Up?" Classroom Computer Capsule, *The College Mathematics Journal*, vol. 27, no. 3, May 1996 and Bradley Efron, "Computer-Intensive Methods in Statistical Regression," *SIAM Review*, vol. 30, no. 3, September 1988.

Chapter 2 Ingredients of Change: Nonlinear Models

2.1 Exponential and Logarithmic Functions and Models

As we begin to consider functions that are not linear, it is very important that you be able to draw scatter plots, find numerical changes in output data, and recognize the underlying shape of the basic functions to be able to identify which function best models a particular set of data. Finding the model is only a means to an end – being able to use mathematics to describe the changes that occur in real-world situations.

2.1.1 ENTERING EVENLY-SPACED INPUT VALUES When an input list consists of many values that are the same distance apart, there is a calculator command that will generate the list so that you do not have to type the values in one by one. The syntax for this sequence command is *seq(formula, variable, first value, last value, increment)*. When entering years that differ by 1, the formula is the same as the variable and the increment is 1. Any letter can be used for the variable -- we choose to use X. We illustrate the sequence command use to enter the input for the population data in Table 2.2 in Section 2.1 of *Calculus Concepts*.

Year	1994	1995	1996	1997	1998	1999	2000	2001	2002	2003
Population	7290	6707	6170	5677	5223	4805	4420	4067	3741	3442

Clear any old data from lists L1 and L2. To enter the input data, position the cursor in the first list so that L1 is darkened.

Generate the list of years beginning with 1994, ending with 2003, and differing by 1 with [2nd] [STAT] (LIST) ▶ [OPS] 5 [seq(] [X,T,θ,n] [,] [X,T,θ,n] [,] 1994 [,] 2003 [,] 1 [)].

Press [ENTER] and the sequence of years is pasted into L1. Manually enter the output values in L2.

2.1.2 FINDING PERCENTAGE DIFFERENCES When the input values are evenly spaced, use program DIFF to compute the percentage differences (also called the *percentage changes*) in the output. If the data are perfectly exponential (that is, every data point falls on the exponential function), the percentage change in the output is constant. If the percentage changes are close to constant, this is an indication that an exponential model *may* be appropriate.

Return to the home screen. Find the percentage change in the population data with program DIFF. Press [PRGM] and then the number that is to the left of the DIFF program location.

Press [ENTER] and the message on the right appears on your screen. Press [STAT] 1 [EDIT] and ▶ ▶ ▶ to view the percentage change in list L5.

A-33

Scroll down L5 with ▼ to see all 9 values. The percentage differences are close to constant, so an exponential function will probably give a good fit to these data.

2.1.3 FINDING AN EXPONENTIAL FUNCTION TO MODEL DATA
Use your calculator to find an exponential equation to model the small town population data. The exponential function we use is of the form $y = ab^x$. It is very important that you align large numbers (such as years) whenever you find an exponential model. If you don't align to smaller values, the TI-83 may return an error message or an incorrect function.

We continue to use the data in Table 2.2 of the text. First, align the input data to smaller values. Other alignments are possible, but we choose to align the data so that $x = 0$ in 1994. To align, darken the list name and press [2nd] 1 (L1) [−] 1994 [ENTER].

Return to the home screen. Following the same procedure that you did to find a linear function, find the exponential function and paste the equation into the Y1 location of the Y= list by pressing [STAT] [▶] [CALC] 0 [ExpReg] [VARS] [▶] [Y-VARS] 1 [Function] 1 [Y1].

Press [ENTER] to find the equation and paste it into the Y1 location. Press [Y=] to view Y1. Press [ZOOM] 9 [ZoomStat] to draw the scatter plot and the graph of the function.

As the percentage changes indicate, the function gives a very good fit for the data. To estimate the town's population in 2004, remember that x is the number of years past 1994 and evaluate the function in Y1 at $x = 10$.
We predict the population to be about 3167 people in 2004.

- Do not confuse the percentages changes (percentage differences) found from the data with the percentage change for the exponential function. The constant percentage change in the exponential function $y = ab^x$ is $(b - 1)100\%$. The constant percentage change for the function is a single value whereas the percentage changes calculated from the data are many different numbers.

2.1.4 FINDING A LOG FUNCTION TO MODEL DATA
Use your calculator to find a log equation of the form $y = a + b \ln x$. We illustrate finding this function with the air pressure and altitude data in Table 2.5 of Section 2.1 of *Calculus Concepts*.

Air pressure (inches of mercury)	13.76	5.56	2.14	0.82	0.33
Altitude (thousands of feet)	20	40	60	80	100

Clear the data that is currently in lists L1 and L2. Enter the air pressure data in L1 and the altitude data in L2. (If you wish, clear list L3, but it is not necessary to do so.)

TI-83 Guide   A-35

Delete (with CLEAR) any functions that are in the Y= list. A scatter plot of the data drawn with ZOOM 9 [ZoomStat] shows the slow decline that can indicate a log model.	L1: 13.76, 5.56, 2.14, .82, .33 L2: 20, 40, 60, 80, 100 L3: ------ L2(6)=	(scatter plot)
As when modeling linear and exponential functions, find and paste the log equation into the Y1 location of the Y= list by pressing STAT ▶ [CALC] 9 [LnReg] VARS ▶ [Y-VARS] 1 [Function] 1 [Y1]. Press Y= to view the pasted function.	Plot1 Plot2 Plot3 \Y1■76.173556931 394+-21.33120171 55331ln(X) \Y2= \Y3= \Y4= \Y5=	
Either press ZOOM 9 [ZoomStat] or GRAPH to view the function on the plot of the data. It appears to be an excellent fit.		(curve with data points)

2.1.5 REVISITING INVERSE FUNCTIONS Recall that exponential and log functions are inverse functions. If you find the inverse function for the log equation that models altitude as a function of air pressure (given in the section above), it will be an exponential function. We briefly explore this concept using the TI-83. Leave the function $y = 76.174 - 21.331 \ln x$ in the Y1 location the Y= list because we use it again in this investigation.

Find the exponential equation that models air pressure as a function of altitude and paste it into the Y2 location with STAT ▶ [CALC] 0 [ExpReg] 2nd 2 (L2) , 2nd 1 (L1) , VARS ▶ [Y-VARS] 1 [Function] 2 [Y2].	ExpReg L2,L1,Y2■ Note that L2 is typed first!	
Press ENTER. Then press Y= and turn Plot 1 off. Enter the composite function Y1∘Y2 (or Y2∘Y1) in the Y3 location. Enter some values in the TABLE and see that what is returned for Y3 is close to the value of X that was entered.	Plot1 Plot2 Plot3 \Y1■76.173556931 394+-21.33120171 55331ln(X) \Y2■35.540860323 521*.95420711463 231^X \Y3■Y1(Y2)■ We use the Composition Property of Inverse Functions.	X: 10, 25, 32 Y2: 22.241, 11.01, 7.9304 Y3: 10.005, 25.004, ■■■■■ Y3=32.0030785833 Press ▶ to see the Y3 column.
You can also graphically verify that Y1 and Y2 are inverse functions. Clear or turn off the composite function(s) and turn off Y2. On the home screen, have the TI-83 draw the graph of the inverse of Y1 with 2nd PRGM (DRAW) 8 (DrawInv) VARS ▶ [Y-VARS] 1 [Function] 1 [Y1] ENTER. Press Y=, turn Y2 on, and press GRAPH. You see the same graph, indicating that Y2 and the inverse function are either the same or close enough to not be visually distinguished.		(graph showing two inverse curves)

Copyright © Houghton Mifflin Company. All rights reserved.

- There is a slight difference in the exponential function and the true inverse function for Y1 because of rounding and because the exponential function in Y2 was not found algebraically but from fitting an equation to data points.

2.2 Exponential Models in Action

In this section we consider further applications to equation solving with the TI-83 as well as its use in problems that deal with the mathematics of finance. You are probably familiar with the compound interest formulas. This section introduces you to some new methods of using your calculator with these familiar formulas.

2.2.1 NOMINAL AND EFFECTIVE INTEREST RATES

The TI-83's FINANCE menu will not help you learn calculus, but it contains many features that might be helpful in other courses. One feature that you can use in this course is the effective – nominal interest rate conversion feature. We illustrate it with the two investment option equations that are in part *a* of Example 1 in Section 2.2 of *Calculus Concepts*:

$$Q(t) = P\left(1 + \frac{0.069}{4}\right)^{4t} \quad \text{and} \quad M(t) = P\left(1 + \frac{0.067}{12}\right)^{12t}$$

The syntax for the TI-83's effective rate is Eff(nominal rate, number of compounding periods per year), and the syntax for the TI-83's nominal rate is Nom(effective rate, number of compounding periods per year). Both rates are entered and returned as percentages, not decimal values.

Find the effective rates for each of the investment options given in part *a* of Example 1. Access the TI-83's effective rate with [2nd] [x^{-1}] (FINANCE) [ALPHA] [PRGM] [▶EFF(].
 	
 We see that the first option with quarterly interest has the higher effective rate and is therefore the better investment.
	▶Eff(6.9,4)
	     7.080599536
	▶Eff(6.7,12)
	     6.909623528

2.2.2 MORE ON SOLVING FOR INPUT VALUES

Solving for a variable that appears in an exponent is the same as solving for a variable that appears elsewhere in an equation when you use the equation solver. We illustrate this idea with the two investment option equations that are in part *b* of Example 1 in Section 2.2 of *Calculus Concepts*: $2P = Q(t)$ and $2P = M(t)$. The functions Q and M are defined in the previous section of this *Guide*.

Remember that when solving an equation on the TI-83, it must be entered in the calculator in the form $f(x) = 0$. If you prefer, you can divide both sides of the two equations by P. We choose to leave P in the equations to illustrate that the value of P does not affect the solutions.

Enter $2P - P\left(1 + \frac{0.069}{4}\right)^{4t}$ in Y1 and $2P - P\left(1 + \frac{0.067}{12}\right)^{12t}$ in Y2. Type *P* with [ALPHA] 8 and replace *t* with X. Be certain to enclose the exponents in parentheses as well as using the parentheses shown in the original formulas.	
Access the SOLVER with [MATH] 0 and clear any previous equations with [▲] [CLEAR]. To solve the equation Y1 = 0, type Y1 in the eqn: location and press [ENTER]. Enter some value for *P* and a guess for X. (Try using different values than those shown to the right to see what happens.)	

TI-83 Guide   A-37

- From this point on, the keystrokes used to access function locations in the Y= list will not be given. You should remember that you must type the function name using [VARS] [▶] [Y–VARS] 1 [Function] followed by the number of the desired function location. The TI-83 does not recognize [ALPHA] 1 (Y) as the name of a function in the Y= list.

> With the cursor on the line containing the variable X, press [ALPHA] [ENTER] (SOLVE).
>
> Repeat the solving procedure using a different value of *P* and a different guess for X to see that you obtain the same answer.

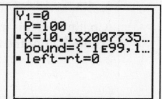

- Why do you obtain the same answer with a different value for *P* and a different guess? Go back to the original equations on the previous page to see that you can divide both sides by *P*, thereby eliminating it from the equations.

- What about the guess? Recall that the solver uses your guess as a starting point to find the answer. If there is more than one answer, the solver returns the one closest to your guess. Here we are dealing with exponential functions on one side of the equation and a linear function (2*P*) on the other side. A line can cross an exponential curve only once, so there is only one answer for the solver to find.

> To solve the second equation, press [▲] until you return to the eqn: screen, change Y1 to Y2, enter a new guess (or leave what is currently there as the guess), be sure the cursor is on the line corresponding to what you are solving for (X), and press [ALPHA] [ENTER] (SOLVE).

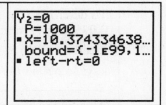

CAUTION: Being able to solve equations is only the first step in obtaining a solution. Always read again the question you are asked and see what you must do with the result from the solver to answer the question. You must also determine if the answer should be rounded due to a discrete interpretation. Your calculator cannot assist you with these tasks, but practice helps a lot!

2.2.3 FINDING FORMULA OUTPUTS BY USING PREVIOUS ENTRIES You can enter a formula in the Y= list and find outputs by either by evaluating the formula at different values on the home screen or by using the ASK feature of the TI-83 TABLE. (See Sections 1.1.5 and 1.1.6.) Another method that you can use is to recall expressions previously typed by repeatedly using the calculator's last entry feature. Often, this method can save time. We illustrate it by using the values in Table 2.23 in Section 2.2 of *Calculus Concepts*.

> On the home screen, type the formula $\left(1+\frac{1}{n}\right)^n$. Type *n*, which prints as a capital letter on the calculator, with [ALPHA] [LOG]. Press [ENTER]. Next, store 1 in *n*.

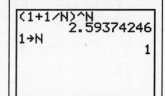

NOTE: The output you obtain depends on the value that is stored for N. You may obtain a different output value than the one shown above because you have a different value stored in N. After storing 1 in N, your next result should be the same as the shown below.

Copyright © Houghton Mifflin Company. All rights reserved.

| Press [2nd] [ENTER] (ENTRY) twice (or as many times as it takes to again display the formula on the screen), and then press [ENTER]. The formula is now evaluated at $n = 1$. Store 2 in $n$ and repeat the process of recalling and evaluating the formula. | ```
(1+1/N)^N
                   1
2→N
                   2
(1+1/N)^N
                2.25
``` |
| Continue in this manner until you find all the amount values in Table 2.23 of the text. | ```
              525600
(1+1/N)^N
         2.718279215
31536000→N
            31536000
(1+1/N)^N
         2.718282473
``` |

2.2.4 DETERMINING FUTURE VALUE

You can often save many keystrokes with the recall of previously-typed expressions by repeatedly using the calculator's last entry feature. When a formula contains more than one input variable, it is easier to recall the last entry on the home screen than to try to use the TABLE.

To illustrate, consider the compound interest formula – one that contains several input variables: Amount $= P\left(1+\frac{r}{n}\right)^{nt}$ dollars is the amount in an account paying $100r\%$ interest (compounded n times a year) on an initial deposit of $\$P$ over a period of t years.

| To determine the amount in an account with $P = \$1500$, $r = 0.075$, $n = 12$, and $t = 1$, first store these values in the memory locations $P$, $R$, $N$, and $T$. (Press [ALPHA] before pressing the keys to obtain the letters $P$, $R$, $N$, and $T$.) | ```
1500→P
                1500
.075→R
                .075
12→N
                  12
1→T■
``` |
| Also on the home screen, type in the formula for the amount in the account – the future value. Remember to use parentheses around the exponent because it consists of more than one symbol. Press [ENTER] to evaluate the formula at the stored values. | ```
                .075
12→N
                  12
1→T
                   1
P(1+R/N)^(NT)
         1616.448898
``` |
| To determine the future value of the account in 2 years, first store 2 in $t$. Press [2nd] [ENTER] (ENTRY) twice (or as many times as it takes to again display the amount formula on the screen), and then press [ENTER]. (The other values remain as you stored them until they are changed.) | ```
2→T
                   2
P(1+R/N)^(NT)
         1741.938027
``` |

2.2.5 FINDING PRESENT VALUE

Suppose you want an alternate method than that indicated in the text to solve the equation in part *a* of Example 5 in Section 2.2 of *Calculus Concepts*:

Solve $9438.40 = P\left(1+\frac{0.075}{12}\right)^{60}$ for the present value P.

| Access the SOLVER with [MATH] 0 and clear any previous equations with [▲] [CLEAR]. To solve this equation, rewrite it so that 0 is on one side and enter the rewritten non-zero side in the eqn: location. | ```
EQUATION SOLVER
eqn:0=9438.4-P(1
+.075/12)^60
``` |

TI-83 Guide A-39

| Press ENTER. Enter a guess for the unknown quantity, P. With the cursor on the line containing the variable P, press ALPHA ENTER (SOLVE). | `9438.4-P(1+.0…=0`
`• P=6494.4858703…`
`  bound={-1E99,1…`
`• left-rt=0` |

2.3 Logistic Functions and Models

This section introduces the logistic function that can be used to describe growth that begins as exponential and then slows down to approach a limiting value. A change in the sign of one of the parameters gives a logistic function that describes decay instead of growth.

Logistic functions are often in applications to the spread of disease. The TI-83's random number generator can be used to collect data to illustrate such a situation.

2.3.1 GENERATING RANDOM NUMBERS Imagine all the real numbers between 0 and 1, including 0 but not 1, written on identical slips of paper and placed in a hat. Close your eyes and draw one slip of paper from the hat. You have just chosen a number "at random."

Your calculator doesn't offer you a random choice of all real numbers between 0 and 1, but it allows you to choose, *with an equal chance of obtaining each one*, any of 10^{14} different numbers between 0 and 1 with its random number generator called rand.

Before using your calculator's random number generator for the first time, you should "seed" the random number generator. (This is like mixing up all the slips of paper in the hat.)

| Pick some number, not the one shown on the right, and store it as the "seed" with *your number* STO▶ MATH ◀ [PRB] 1 [rand] ENTER. Everyone needs to have a different seed, or the number choices will not be random. | `2658→rand`
`            2658` |

| Enter rand again, and press ENTER several times.

Your list of random numbers should be different from the one on the right if you entered a different seed. Notice that all numbers are between 0 and 1. | `rand`
`    .7005331056`
`    .9836308181`
`    .0068461306`
`    .2110053651`
`    .2225253266`
`    .728178859` |

We usually count with the numbers 1, 2, 3, ... and so forth, not with decimal values. So, to adapt the random number generator to choose a whole number between, and including, 1 and N, we need to enter the instruction int(N rand + 1).

| Enter, using a specific number for N, MATH ▶ 5 [int(] N MATH ◀ [PRB] 1 [rand] + 1). For instance, if you want to choose a person at random from a room that contains 10 people, use N = 10.

Repeatedly press ENTER to choose more random numbers. | `int(10rand+1)`
`            7`
`            8`
`            4`
`            6`
`           10`
`           10` |

As, previously stated, your list of numbers should be different from those shown above. All values will be between 1 and 10. Note that it is possible to obtain the same value more than once. This corresponds to each slip of paper being put back in the hat after being chosen and the numbers mixed well before the next number is drawn.

2.3.2 FINDING A LOGISTIC FUNCTION TO MODEL DATA

You can use your calculator to find a logistic function of the form $y = \dfrac{L}{1+Ae^{-Bx}}$. Unlike the other functions we have fit to data, the logistic equation that you obtain may be slightly different from a logistic equation that is found with another calculator. (Logistic equations in *Calculus Concepts* were found using a TI-83.) As you did when finding an exponential equation for data, large input values must be aligned or an error or possibly an incorrect answer could be the result.

Note that the TI-83 finds a "best-fit" logistic function rather than a logistic function with a limiting value L such that no data value is ever greater than L. We illustrate finding a logistic function using the data in Example 1 of Section 2.3 in *Calculus Concepts*.

End of month	Jan	Feb	Mar	Apr	May	June	July	Aug	Sept
Total number of swimsuits	4	12	25	58	230	439	648	748	769

Clear any old data. Delete any functions in the Y= list and turn on Plot 1. Using an input of 1 for January, 2 for February, and so forth, enter the data in the above table in lists L1 and L2:

Construct a scatter plot of the data. An inflection point is indicated, and the data appear to approach limiting values on the left and on the right. A logistic model seems appropriate.	
To find the logistic function and paste the equation into the Y1 location of the Y= list, press [STAT] [▶] [CALC] [ALPHA] B [Logistic] [VARS] [▶] [Y-VARS] 1 [Function] 1 [Y1].	

Note: TI-83 menu items can be chosen by using [▼] to scroll to the location of the item and pressing [ENTER]. For instance, instead of choosing Logistic by pressing [ALPHA] B as indicated in the instruction above, you can scroll down the list of functions until B is highlighted and press [ENTER].)

Press [ENTER]. This equation will take longer to generate than the other equations. Note that the TI-83 uses the variable c for the value that the text calls L.	
Press [GRAPH] (or [ZOOM] 9 [ZoomFit]) to see the graph of the function drawn on the scatter plot.	
The limiting value, in this context, is about 787 swimsuits. The two horizontal asymptotes are $y = 0$ (the line lying along the x-axis) and the line $y \approx 787$. To see the upper asymptote drawn on a graph of the data, enter 787 in Y2 and press [GRAPH].	

- Provided the input values are evenly spaced, program DIFF might be helpful when you are trying to determine if a logistic model is appropriate for certain data. If the first differences (in list L3 after running program DIFF) begin small, peak in the middle, and end small, this is an indication that a logistic model may provide a good fit to the data. Such is true for this data set because the first differences are 8, 13, 33, 172, 209, 209, 100, and 21.

2.3.3 RECALLING MODEL PARAMETERS Rounding function parameters can often lead to incorrect or misleading results. You may find that you need to use the complete values of the coefficients after you have found a function that best fits a set of data. It would be tedious to copy all these digits in a long decimal number into another location of your calculator. You don't have to! The following procedure applies for any function that you find using one of the built-in regressions (*i.e.*, a function from the STAT CALC menu) in your TI-83.

We illustrate these ideas by recalling the parameter c for the logistic function found in the previous section. To recall the value of c in $y = c/(1 + ae^{\wedge}(-bx))$, press $\boxed{\text{VARS}}$ 5 [Statistics] $\boxed{\blacktriangleright}$ $\boxed{\blacktriangleright}$ [EQ] 4 (c).	XY Σ EQ TEST PTS 1:RegEQ 2:a 3:b 4:c 5:d 6:e 7↓r
Press $\boxed{\text{ENTER}}$ and the full value of c will be "pasted" where you had the cursor before pressing the keys that call up the value of c. (If you did not obtain the correct value for c, repeat finding the logistic function in the previous section and then repeat the previous instructions to find c.)	c 786.7044459

- This same procedure applies to any equation you have found using the STAT CALC menu. For instance, if you had just fit an exponential function to data, the values of a and b in the EQ menu would be for that function and the other parameters would not apply.

WARNING: Once a different function is found, no parameters from a previous function fit to data remain stored in the calculator's memory.

2.4 Polynomial Functions and Models

You will in this section learn how to fit functions that have the familiar shape of a parabola or a cubic to data. Using your calculator to find these equations involves the same procedure as when using it to find linear, exponential, log, or logistic functions.

2.4.1 FINDING SECOND DIFFERENCES When the input values are evenly spaced, you can use program DIFF to quickly compute second differences in the output values. If the data are perfectly quadratic (that is, every data point falls on a quadratic function), the second differences in the output values are constant. When the second differences are close to constant, a quadratic function *may* be appropriate for the data.

We illustrate these ideas with the roofing jobs data given at the beginning of Section 2.4 in *Calculus Concepts*. We align the input data so that 1 = January, 2 = February, etc. Clear any old data and enter these data in lists L1 and L2:

Month of the year	1	2	3	4	5	6
Number of roofing jobs	12	14	22	37	58	84

The input values are evenly spaced, so we can see what information is given by viewing the second differences.

Run program DIFF and observe the first differences in list L3, the second differences in L4, and the percentage differences in list L5.

The second differences are close to constant, so a quadratic function may give a good fit for these data.

Construct a scatter plot of the data. (Don't forget to clear the Y= list of previously-used equations and to turn on Plot 1.)

The shape of the data confirms the numerical investigation result that a quadratic function is appropriate.

2.4.2 FINDING A QUADRATIC FUNCTION TO MODEL DATA Use your TI-83 to find a quadratic function of the form $y = ax^2 + bx + c$ to model the roofing jobs data.

Find the quadratic function and paste the equation into the Y1 location of the Y= list by pressing [STAT] [▶] [CALC] 5 [QuadReg] [VARS] [▶] [Y-VARS] 1 [Function] 1 [Y1] [ENTER].

Press [GRAPH] to overdraw the graph of the function on the scatter plot. This function gives a very good fit to the data.

How many jobs do we predict the company will have in August? Because August corresponds to $x = 8$, evaluate Y1 at $x = 8$. We predict that there will be 156 jobs in August.

2.4.3 FINDING A CUBIC FUNCTION TO MODEL DATA Whenever a scatter plot of data shows a single change in concavity, we are limited to fitting either a cubic or logistic function. If one or two limiting values are apparent, use the logistic equation. Otherwise, a cubic function should be considered. When appropriate, use your calculator to obtain the cubic function of the form $y = ax^3 + bx^2 + cx + d$ that best fits a set of data.

We illustrate finding a cubic function with the data in Table 2.44 in Example 3 of Section 2.4 in *Calculus Concepts*. The data give the average price in dollars per 1000 cubic feet of natural gas for residential use in the U.S. for selected years between 1980 and 1997.

Year	1980	1982	1985	1990	1995	1996	1997
Price (dollars)	3.68	5.17	6.12	5.77	6.06	6.34	6.93

First, clear your lists, and then enter the data. Next, align the input data so that x represents the number of years since 1980.

(We do not have to align here, but we do so in order to have smaller coefficients in the cubic function.)

Draw a scatter plot of these data. Notice that a concavity change is evident, but there do not appear to be any limiting values. Thus, a cubic model is appropriate to fit the data.	
Find the cubic function and paste the equation in Y1 with [STAT] [▶] [CALC] 6 [CubicReg] [VARS] [▶] [Y–VARS] 1 [Function] 1 [Y1] [ENTER].	CubicReg y=ax³+bx²+cx+d a=.0042652511 b=-.1181629878 c=.967358642 d=3.68097369
Overdraw the graph of the function on the scatter plot with [GRAPH] or [ZOOM] 9 [ZoomFit].	
What was the price of 1000 cubic feet of natural gas in 1994? Because x represents the number of years after 1980, evaluate Y1 at $x = 14$ to find that the answer is \$5.77.	Y1(14) 5.767897956

According to the model, when did the price of 1000 cubic feet of natural gas first exceed \$6? You can find the answer by using the SOLVER to find the solution to Y1 – 6 = 0. However, because we already have a graph of this function, the following may be easier. Note that if you do use the SOLVER, you should first TRACE the graph to estimate a guess for x that you need to enter in the SOLVER.

Press [Y=] and enter 6 in Y2. Press [GRAPH]. Note that there are 3 solutions to the equation Y1 = 6. We are asked to find the one that occurs first. Press [2nd] [TRACE] [CALC] 5 [intersect], press [ENTER] [ENTER] to choose the 2 functions, and use [◀] to move the cursor near the first intersection point.	Y2=6 Guess? X=4.593617 Y=6
Press [ENTER] to find the input at the point of intersection. Because x is the number of years after 1980, the answer to the question posed in part c of Example 3 is either 1984 or 1985.	Intersection X=4.3274186 Y=6

The price had not exceeded \$6 in 1984, so the answer to the question is 1985. Carefully read the question you are asked!	Y1(4)
	    5.93277652
	Y1(5)
	    6.096848585

2.5 Shifting Data

If a scatter plot of data, numerical differences calculated from data, and/or a particular context indicate that a certain function is appropriate, but the function does not appear to fit the data, it may be that a horizontal or vertical shift of the data should be considered.

2.5.1 SHIFTING DATA VERTICALLY

We illustrate a vertical shift to improve the fit using the investment club data in Table 2.66 in Example 1 of Section 2.5 of *Calculus Concepts*:

Year	1990	1991	1992	1993	1994	1995	1996	1997	1998
Number of clubs	7085	7360	8267	10,033	12,429	16,054	25,409	31,828	36,500

First, clear your lists, and then enter the data in the table. Next, align the input data so that x represents the number of years since 1990. Draw a scatter plot of the data. The context of the situation and a scatter plot of the data indicate that a logistic model may be appropriate. (Note that it is necessary to align the input data to smaller values because of the exponential term in the logistic function.)

However, when we find a logistic function, it does not fit the data very well – for instance, the concavity of the data changes but the concavity of the function does not. Also, the limiting value in the logistic equation (about 244,096.9318) is much larger than what the data indicates.	
This situation calls for shifting the data vertically so that it approaches a lower asymptote of $y = 0$, which is what the logistic function in the TI-83 has as its lower asymptote for an increasing logistic curve. There are many choices of how to shift the data. We choose to subtract 7000 from each output data value.	
Press [ENTER] to shift the output data. Now find a logistic function to fit the aligned and shifted data. Note that you can return to the home screen and press [2nd] [ENTER] (ENTRY) to repeat the last instruction "Logistic Y1". Press [ENTER].	
This shifted function is a much better fit for the aligned shifted data. Press [ZOOM] 9 [ZoomFit] to see the scatter plot with the new function overdrawn.	
We were to find a function to fit the original, not the shifted, data. Because we subtracted 7000 from each data value, we must add 7000 to the function for the shifted data. Press [Y=] and use [▶] and [▼] to move to the end of the function. Add 7000.	

If you wish to graph this function on the original data, add 7000 to each output value by having the list name L2 darkened and pressing [2nd] 2 (L2) [+] 7000 [ENTER] and draw the graphs with [ZOOM] 9 [ZoomFit]. If you are unsure which of two shifted logistic functions better fits the data, zoom in on each near the "ends" of each graph to see which gives the better fit.

Chapter 3   Describing Change: Rates

3.1 Change, Percent Change, and Average Rates of Change

As you calculate average and other rates of change, remember that every numerical answer in a context should be accompanied by units telling how the quantity is measured. You should also be able to interpret each numerical answer. It is only through their interpretations that the results of your calculations will be useful in real-world situations.

3.1.1 FINDING AVERAGE RATE OF CHANGE   Finding an average rate of change using a function is just a matter of evaluating the equation at two different values of the input variable and dividing by the difference of those input values.

We illustrate this concept using the function describing the population density of Nevada from 1950 through 1997 that is given in Example 3 of Section 3.1 of *Calculus Concepts*.

> The population density referred to above is given by the function $p(t) = 0.1617(1.04814^t)$ people per square mile where t is the number of years after 1900. Press [Y=], clear any functions, turn Plot 1 off, and enter this function in Y1 using X as the input variable.

To find the average rate of change of the population density from 1950 through 1980, first realize that 1950 corresponds to $x = 50$ and 1980 corresponds to $x = 80$. Then, recall that the average rate of change of p between 1950 and 1980 is given by the quotient $\frac{p(80)-p(50)}{80-50}$.

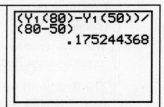

> Return to the home screen and type the quotient $\frac{Y1(80)-Y1(50)}{80-50}$.
> Remember to enclose both the numerator and denominator of the fraction in parentheses.
>
> Finding this quotient in a single step avoids having to round intermediate calculation results.

Recall that rate of change units are output units per input units. On average, Nevada's population density increased about 0.18 people per square mile per year between 1950 and 1980.

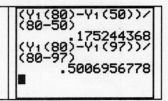

> To find the average rate of change between 1980 and 1997, recall the last expression with [2nd] [ENTER] (ENTRY) and type over 50 with 97 in two places. Press [ENTER].

- If you have many average rates of change to calculate, you could put the average rate of change formula in the graphing list: Y2 = (Y1(B) – Y1(A))/(B – A). Of course, you need to have a function in Y1. Then, on the home screen, store the inputs of the two points in *A* and *B* with 80 [STO▸] [ALPHA] A [ALPHA] [.] (:) 90 [STO▸] [ALPHA] B [ENTER]. All you need do next is type Y2 and press [ENTER]. Then, store the next set of inputs into *A* and recall Y2, using [2nd] [ENTER] (ENTRY) to recall each instruction, press [ENTER], and you have the average rate of change between the two new points. Try it!

3.1.2 CALCULATING PERCENTAGE CHANGE You can find percentage changes using data either by the formula or by using program DIFF. To find a percentage change from a function instead of data, you should use the percentage change formula. We again use the population density of Nevada function in Example 3 of Section 3.1 of *Calculus Concepts* to illustrate.

| Have $p(x) = 0.1617(1.04814^x)$ in Y1. The percentage change in the population density between 1950 and 1997 is given by the formula $\frac{p(97)-p(50)}{p(50)} \cdot 100\% = \frac{Y1(97)-Y1(50)}{Y1(50)} \cdot 100\%$. Recall that if you type in the entire quotient, you avoid rounding errors. | `(Y1(97)-Y1(50))/`
`Y1(50)`
`        8.113954895`
`Ans*100`
`        811.3954895` |

The population density of Nevada increased by about 811.4% between 1950 and 1997.

3.2 Instantaneous Rates of Change

We first examine the principle of local linearity, which says that if you are close enough, the tangent line and the curve are indistinguishable. We also explore two methods for using the calculator to draw a tangent line at a point on a curve.

3.2.1 MAGNIFYING A PORTION OF A GRAPH The ZOOM menu of your calculator allows you to magnify any portion of a graph. Consider the graph shown in Figure 3.7 in Section 3.2 of *Calculus Concepts*. The temperature model is $T(x) = -0.804x^2 + 11.644x + 38.114$ degrees Fahrenheit where x is the number of hours after 6 a.m.

Enter the temperature equation in Y1 and draw the graph between 6 a.m. ($x = 0$) and 6 p.m. ($x = 12$) using ZOOM ▲ [ZoomFit]. We now want to zoom and "box" in several points on the graph to see a magnified view at those points.	
The first point we consider on the graph is point A with $x = 3$. Press ZOOM 1 [ZBox] and use the arrow keys (◄, ▲, etc.) to move the cursor to the left of the curve close to where $x = 3$. (You may not have the same coordinates as those shown on the right.) Press ENTER to fix the lower left corner of the box.	X=2.5531915 Y=65.312586
Use the arrow keys to move the cursor to the opposite corner of your "zoom" box. Point A should be close to the center of your box.	X=3.8297872 Y=69.392374
Press ENTER to magnify the portion of the graph that is inside the box. Look at the dimensions of the view you now see with WINDOW. Repeat the above process if necessary. The closer you zoom in on the graph, the more it looks like a line.	

Reset Xmin to 0, Xmax to 12, and redraw the graph of T with ZoomFit. Repeat the zoom and box steps for point B with $x = 7.24$ and point C with $x = 9$. Note for each point how the graph of the parabola looks more like a line the more you magnify the view. These are illustrations of *local linearity*.

3.2.2 DRAWING A TANGENT LINE

The DRAW menu of the TI-83 contains the instruction to draw a tangent line to the graph of a function at a point. To illustrate the process, we draw several tangent lines on the graph of T, the temperature function given in the previous section of this *Guide*.

Enter $T(x) = -0.804x^2 + 11.644x + 38.114$ in Y1. Next, press WINDOW and set Xmin = 0, Xmax = 12, Ymin = 38, and Ymax = 85. Press GRAPH. With the graph on the screen, press 2nd PRGM (DRAW) 5 [Tangent(], then press 3 ENTER, and the tangent line is drawn at $x = 3$.

Note that the equation of the tangent line and the input value of the point of tangency are displayed at the bottom of the screen. Compare this line with the zoomed-in view of the curve at point A (in the previous section). They actually are the same line if you are at point A!

We illustrate a slightly different method to draw the tangent line to the curve at point B where $x = 7.24$.

Return to the home screen and press 2nd PRGM (DRAW) 5 [Tangent(] VARS ▶ [Y-VARS] 1 [Function] 1 [Y1] , 7.24).

Press ENTER and the tangent line is drawn at the high point of the parabola. Compare this line with the zoomed-in view of the curve at point B (in the previous section).

Note that this method of drawing the tangent line does not show the equation of the tangent nor the point of tangency.

Choose the method you prefer from the two methods given above and draw the tangent at point C where $x = 9$.

(Compare your graph with the one shown in Figure 3.9 in Section 3.2 of the text.)

3.2.3 VISUALIZING THE LIMITING PROCESS

This section of the *Guide* is optional, but it might help you understand what it means when we say that the tangent line is the limiting position of secant lines. Program SECTAN can be used to view secant lines between a point $(a, f(a))$ and close points on a curve $y = f(x)$. The program also draws the tangent line at the point $(a, f(a))$. Before using program SECTAN, a function (using x as the input variable) must be entered in the Y1 location of the Y= list, and you must draw a graph of the function. We illustrate with the graph of the function T that was used in the previous two sections, but you can use this program with any graph. (Program SECTAN is in the *TI-83 Program Appendix*.)

Caution: In order to properly view the secant lines and the tangent line, it is essential that you first draw a graph of the function clearly showing the function, the point of tangency (which should be near the center of the graph), and enough space so that the close points on either side can be seen.

Have $T(x) = -0.804x^2 + 11.644x + 38.114$ in Y1. Next, press WINDOW and set Xmin = 0, Xmax = 7, Ymin = 30, and Ymax = 90. Press GRAPH. Press PRGM and the number or letter corresponding to the location of program SECTAN. (Your program list may not look like the one on the right.)	EDIT NEW 1:DIFF 2:EULER 3:LOGISTIC 4:LSLINE 5:NUMINTGL 6:SECTAN 7:SINREG
Press ENTER to start the program. If you forgot to enter the function in Y1 or to draw its graph, enter 2. Otherwise, type 1 and press ENTER. At the next prompt, type the input value at the point of tangency. (For this illustration, choose $x = 4$.)	HAVE F(X)IN Y1... DRAW GRAPH OF F CONTINUE? YES(1) NO(2)1 X-VALUE OF POINT OF TANGENCY? 4
Press ENTER. The next message that appears tells you to press enter to see secant lines drawn between the point of tangency and close points to the *left*. Press ENTER. (Five secant lines will draw.)	
When you finish looking at the graph, press ENTER to continue. The next message that appears tells you to press enter to see secant lines drawn between the point of tangency and close points to the *right*. (Again, five secant lines will draw.) Press ENTER.	
Press ENTER to continue the program. You are next told to press ENTER to see a graph of the line that is tangent to the curve at $x = 4$.	

As you watch the graphs, you should notice that the secant lines are becoming closer and closer to the tangent line as the close point moves closer and closer to the point of tangency.

3.2.4 TANGENT LINES AND INSTANTANEOUS RATES OF CHANGE

Sometimes the TI-83 gives results that are not the same as the mathematical results you expect. This does not mean that the calculator is incorrect – it does, however, mean that the calculator programming is using a different formula or definition than the one that you are using. You need to know when the TI-83 produces a different result from what you expect.

In this section, we investigate what the calculator does if you ask it to draw a tangent line where the line cannot be drawn. Consider these special cases:

1. What happens if the tangent line is vertical? We consider the function $f(x) = (x + 1)^{1/3}$ which has a vertical tangent at $x = -1$.

2. How does the calculator respond when the instantaneous rate of change at a point does not exist? We illustrate with $g(x) = |x| - 1$, a function that has a sharp point at $(0, -1)$.

3. Does the calculator draw the tangent line at the break point(s) of a piecewise continuous function? We consider two situations:
   a. $h(x)$, a piecewise continuous function that is continuous at all points, and
   b. $m(x)$, a piecewise continuous function that is not continuous at $x = 1$.

TI-83 Guide

1. Enter the function $f(x) = (x + 1)^{1/3}$ in the Y1 location of the Y= list. Remember that anytime there is more than one symbol in an exponent and you are not sure of the TI-83's order of operations, enclose the power in parentheses.

   Draw the graph of the function with ZOOM 4 [ZDecimal]. With the graph on the screen, press 2nd PRGM (DRAW) 5 [Tangent(], then press (-) 1 ENTER. The tangent line is drawn at $x = -1$.

CAUTION: Note that the TI-83 correctly draws the vertical tangent. However, the equation of the vertical tangent line is $x = -1$, <u>not</u> $y = 100x + 100$! The line whose equation is printed on the calculator screen has slope 100, but the instantaneous rate of change at $(-1, 0)$ does not exist because the tangent is vertical.

2. Clear Y1 and enter the function $g(x) = |x| - 1$. The absolute value symbol is typed with MATH ▶ 1 [abs(].

   Draw the graph of the function with ZOOM 4 [ZDecimal]. With the graph on the screen, press 2nd PRGM (DRAW) 5 [Tangent(], then press ENTER. The tangent line is drawn at $x = 0$.

THIS IS INCORRECT! There is a sharp point on the graph of g at $(0, -1)$, and the limiting positions of secant lines from the left and the right at that point are different. A tangent line cannot be drawn on the graph of g at $(0, -1)$ according to *our* definition of instantaneous rate of change. The TI-83's definition is different, and this is why the line is drawn. (The TI-83 definition is explained in Section 4.3.1 of this *Guide*.)

CAUTION: Note that the TI-83 gives the correct equation for the line that is drawn. However, <u>there is no tangent line at this sharp point</u> on this continuous function.

3a. Clear Y1 and Y2 and enter, as indicated, the function
$$h(x) = \begin{cases} x^2 & \text{when } x \leq 1 \\ x & \text{when } x > 1 \end{cases}$$
Recall that the inequality symbols are accessed with 2nd MATH (TEST).

Set each part of the function to draw in DOT mode by placing the cursor over the equals sign in each function, pressing ◀, and then pressing ENTER 6 times until the slanted line turns to a dotted line.

Draw the graph of the function with ZOOM 4 [ZDecimal]. With the graph on the screen, press 2nd PRGM (DRAW) 5 [Tangent(], then press 1 and ENTER. The tangent line is drawn at the break point, $x = 1$.

THE GRAPH YOU SEE IS INCORRECT! Even though h is continuous for all values of x, there is no tangent line to the graph of h at $x = 1$. Secant lines drawn using close points on the right and on the left of $x = 1$ do not approach the same slope, so the instantaneous rate of change of h does not exist at $x = 1$.

NOTE: There is no tangent line at the break point of a piecewise continuous function (even if that function is continuous at the break point) unless secant lines drawn through close points to the left and right of that point approach the same value.

3b. Edit Y2 by placing the cursor over the right parenthesis following X and pressing 2nd DEL (INS) + 1 to enter

$$m(x) = \begin{cases} x^2 & \text{when } x \leq 1 \\ x+1 & \text{when } x > 1 \end{cases}$$

The TI-83 should still be in DOT mode from the previous graph.

Draw the graph of the function with ZOOM 4 [ZDecimal]. With the graph on the screen, press 2nd PRGM (DRAW) 5 [Tangent(], then press 1 and ENTER. The tangent line is drawn at the break point, $x = 1$.

CAUTION: THE GRAPH YOU SEE IS INCORRECT! Because m is not continuous when $x = 1$, the instantaneous rate of change does not exist at that point. The tangent line cannot be drawn on the graph of m when $x = 1$!

Be certain that the instantaneous rate of change exists at a point and the tangent line exists at that point before using your calculator to draw a tangent line or use the tangent equation that may be printed on the screen. Because of the method that your calculator uses to compute instantaneous rates of change, it might draw a tangent line at a point on a curve where the tangent line, according to our definition, does not exist.

3.3 Derivatives

There are no new calculator techniques in this section, but we illustrate a new calculation.

3.3.1 CALCULATING PERCENTAGE RATE OF CHANGE
Suppose the growth rate of a population is 50,000 people per year and the current population size is 200,000 people.

What is the percentage rate of change of the population? The answer is 25% per year.	50000/200000
	          .25
	Ans*100
	          25

Suppose instead that the current population size is 2 million. What is the percentage rate of change? The answer is 2.5% per year, which is a much smaller percentage rate of change.	50000/(2*10^6) .025 Ans*100

Chapter 4 Determining Change: Derivatives

4.1 Numerically Finding Slopes

Using your calculator to find slopes of tangent lines does not involve a new procedure. However, the techniques that are discussed in this section allow you to repeatedly apply a method of finding slopes that gives quick and accurate results.

4.1.1 NUMERICALLY ESTIMATING SLOPES ON THE HOME SCREEN

Finding the slopes of secant lines joining the point at which the tangent line is drawn to increasingly close points on a function to the left and right of the point of tangency is easily done using your TI-83. Suppose we want to numerically estimate the slope of the tangent line at $t = 8$ to the graph of the function that gives the number of polio cases in 1949: $y = \dfrac{42{,}183.911}{1 + 21{,}484.253 e^{-1.248911 t}}$ where $t = 1$ on January 31, 1949, $t = 2$ on February 28, 1949, and so forth.

| Enter the polio cases equation in the Y1 location of the Y= list. (Carefully check your entry of the equation, especially the location of the parentheses.)<br><br>We now evaluate the slopes of secant lines that join close points to the left of $x = 8$ with $x = 8$. | ```
Plot1 Plot2 Plot3
\Y1=42183.911/(1
+21484.253e^(-1.
248911X))
\Y2=
\Y3=
\Y4=
\Y5=
``` |
| On the home screen, type in the expression shown to the right to compute the slope of the secant line joining the close point where $x = 7.9$ and the point of tangency where $x = 8$.<br><br>Record on paper each slope, to at least 1 more decimal place than the desired accuracy, as it is computed. You are asked to find the nearest whole number that these slopes are approaching, so record at least one decimal place in your table of slopes. | ```
(Y1(8)-Y1(7.9))/
(8-7.9)
        13159.68272
``` |
| Press [2nd] [ENTER] (ENTRY) to recall the last entry, and then use the arrow keys to move the cursor over the 9 in the "7.9". Press [2nd] [DEL] (INS) and press 9 to insert another 9 in <u>both</u> places that 7.9 appears. Press [ENTER] to find the slope of the secant line joining $x = 7.99$ and $x = 8$. | ```
(Y1(8)-Y1(7.9))/
(8-7.9)
        13159.68272
(Y1(8)-Y1(7.99))
/(8-7.99)
        13170.61766
``` |
| Continue in this manner, recording each result on paper, until you can determine to which value the slopes from the left seem to be getting closer and closer.<br><br>It appears that the slopes of the secant lines from the left are approaching 13,170 cases per month. | ```
        13170.18713
(Y1(8)-Y1(7.9999
))/(8-7.9999)
        13170.12882
(Y1(8)-Y1(7.9999
9))/(8-7.99999)
        13170.1229
``` |
| We now evaluate the slopes joining the point of tangency and nearby close points to the *right* of $x = 8$.<br><br>Clear the screen, recall the last expression with [2nd] [ENTER] (ENTRY), and edit it so that the nearby point is $x = 8.1$. Press [ENTER] to calculate the secant line slope. | ```
(Y1(8)-Y1(8.1))/
(8-8.1)
        13146.38421
``` |
```

Continue in this manner as you did when calculating slopes to the left, but each time insert a 0 before the "1" in two places in the close point. Record each result on paper until you can determine the value the slopes from the right are approaching.

It appears that the slopes of the secant lines from the right are approaching 13,170 cases per month.

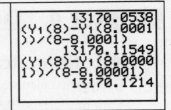

When the slopes from the left and the slopes from the right approach the same number, that number is the slope of the tangent line at the point of tangency. In this case, we estimate the slope of the tangent line to be 13,170 cases per month.

4.1.2 NUMERICALLY ESTIMATING SLOPES USING THE TABLE The process discussed in Section 4.1.1 can be done in fewer steps and with fewer keystrokes when you use the TI-83 TABLE. The point of tangency is $x = 8$, $y = Y1(8)$, and let's call the close point $(x, Y1(x))$. Then, slope = $\frac{\text{rise}}{\text{run}} = \frac{Y1(X) - Y1(8)}{X - 8}$. We illustrate numerically estimating the slope using the TABLE with the logistic function given in Section 4.1.1 of this *Guide*.

Have the polio cases equation given in the previous section in the Y1 location of the Y= list. Enter the above slope formula in another location, say Y2. (Also remember to enclose both the numerator and denominator of the slope formula in parentheses.)

Turn off Y1 because we are only considering the output from Y2.

Press [2nd] [WINDOW] (TBLSET) and choose ASK in the Indpnt: location. (See page A-7 for more specific instructions.)

Access the table with [2nd] [GRAPH] (TABLE) and delete or type over any previous entries in the X column. Enter values for X, the input of the close point, so that X gets closer and closer to 8 from the left.

- Notice that as you continue to enter numbers, the calculator switches the input values you enter to scientific notation and displays rounded output values so that the numbers can fit on the screen in the space allotted for outputs of the table. You should position the cursor over each output value and record on paper as many decimal places as necessary in order to determine the limit from the left to the desired degree of accuracy.

Repeat the process, entering values for X, the input of the close point, so that X gets closer and closer to 8 from the right.

Remember to highlight the output values and record on paper the slope to at least one more decimal position than the desired accuracy.

The limit is estimated to be the value that the limits from the left and the right are approaching: 13,170 cases per month.

Note: You may wish to leave the slope formula in Y2 as long as you need it. Turn Y2 off when you are not using it.

TI-83 Guide

4.2 Algebraically Finding Slopes

The TI-83 does not find algebraic formulas for slope, but you can use the built-in numerical derivative and draw the graph of a derivative to check any formula that you find algebraically.

4.2.1 UNDERSTANDING YOUR CALCULATOR'S SLOPE FUNCTION The TI-83 uses the slope of a secant line to approximate the slope of the tangent line at a point on the graph of a function. However, instead of using a secant line through the point of tangency and a close point, the TI-83 uses the slope of a secant line through two close points that are equally spaced from the point of tangency.

The secant line joining the points $(a - k, f(a - k))$ and $(a + k, f(a + k))$ and the line tangent to the graph of f where $x = a$ are shown in Figure 4. Notice that the slopes of the secant line and tangent appear to be close to the same value even though these are different lines.

Figure 4

As k gets closer and closer to 0, the two points through which the secant line passes move closer and closer to a. Provided the slope of the tangent line exists, the limiting position of the secant line will be the tangent line.

The calculator's notation for the slope of the secant line shown in Figure 4 is

nDeriv(function, symbol for input variable, a, k)

Specifying the value of k is optional. If a value for k is not given, the calculator automatically uses $k = 0.001$. Any smooth, continuous function will do, so let's investigate these ideas with the function in Example 3 of Section 4.2 of *Calculus Concepts*: $f(x) = 2\sqrt{x}$.

Enter $f(x) = 2\sqrt{x}$ in one of the locations of the Y= list, say Y1. Return to the home screen with [2nd] [MODE] (QUIT).	Plot1 Plot2 Plot3 \Y1■2√(X) \Y2= \Y3= \Y4= \Y5= \Y6= \Y7=
Suppose you want to find the slope of the secant line between the points $(0, f(0))$ and $(2, f(2))$. That is, you are finding the slope of the secant line between the points $(a - k, f(a - k))$ and $(a + k, f(a + k))$ for $a = 1$ and $k = 1$. Type the expression on the right and then press [ENTER]. Access nDeriv(with [MATH] 8 [nDeriv(].	nDeriv(Y1,X,1,1) 1.414213562

Remember that when you are on the home screen, you can recall previous instructions with the keystrokes [2nd] [ENTER] (ENTRY).

Copyright © Houghton Mifflin Company. All rights reserved.

Recall the last entry and edit the expression so that k changes from 1 to 0.1. Press ENTER. Again recall the last entry, and edit the expression so that k changes from 0.1 to 0.01. Press ENTER.	1.414213562
	nDeriv(Y1,X,1,.1)
	        1.001255501
	nDeriv(Y1,X,1,.01)
	        1.000012501

Repeat the process for $k = 0.001$ and $k = 0.0001$. Note how the results are changing. As k becomes smaller and smaller, the secant line slope is becoming closer and closer to 1.	1.000012501
	nDeriv(Y1,X,1,.001)
	        1.000000125
	nDeriv(Y1,X,1,.0001)
	        1.000000002

- A logical conclusion is that the tangent line slope is 1. However, realize that we have just done another type of numerical investigation, not an algebraic proof.

In the table below, the first row lists some values of a, the input of a point of tangency, and the second row gives the actual slope (to 7 decimal places) of the tangent line at those values. The algebraic method gives the exact slope of the line tangent to the graph of f at these input values.

Use your calculator to verify the values in the third through sixth rows that give the slope of the secant line (to 7 decimal places) between the points $(a - k, f(a - k))$ and $(a + k, f(a + k))$ for the indicated values of k. Find each secant line slope with nDeriv(Y1, X, a, k).

a = input of point of tangency	2.3	5	12.82	62.7
slope of tangent line = $f'(a)$	0.6593805	0.4472136	0.2792904	0.1262892
slope of secant line, $k = 0.1$	0.6595364	0.4472360	0.2792925	0.1262892
slope of secant line, $k = 0.01$	0.6593820	0.4472138	0.2792904	0.1262892
slope of secant line, $k = 0.001$	0.6593805	0.4472136	0.2792904	0.1262892
slope of secant line, $k = 0.0001$	0.6593805	0.4472136	0.2792904	0.1262892

You can see that the slope of the secant line is very close to the slope of the tangent line for small values of k. For this function, the slope of the secant line does a great job of approximating the slope of the tangent line when k is very small.

Now repeat the process, but do not include k in the instruction. That is, find the secant line slope by calculating nDeriv(Y1, X, a, k). Did you obtain the following (to 7 decimal places)?

slope of secant line	0.6593805	0.4472136	0.2792904	0.1262892

These values are those in the fifth row of the above table – the values for $k = 0.001$. From this point forward, we use $k = 0.001$ and therefore do not specify k when evaluating nDeriv(. Will the slope of this secant line always do a good job of approximating the slope of the tangent line when $k = 0.001$? Yes, it generally does, as long as the instantaneous rate of change exists at the input value at which you evaluate nDeriv(. When the instantaneous rate of change does not exist, nDeriv(should *not* be used to approximate something that does not have a value!

You will benefit from reading again Section 3.2.4 of this *Guide*, which illustrates several cases when the TI-83's slope function gives a value for the slope when it does not exist. For instance, the instantaneous rate of change of $f(x) = |x|$ does not exist at $x = 0$ because the graph of f has a sharp point there. However, the TI-83's slope function does exist at that point!	

TI-83 Guide

4.4 More Simple Rate-of-Change Formulas

The TI-83 only approximates numerical values of slopes – it does not give a slope in formula form. You also need to remember that the TI-83 calculates the slope (*i.e.*, the derivative) at a specific input value by a different method than the one we use to calculate the slope.

4.4.1 DERIVATIVE NOTATION AND CALCULATOR NOTATION In addition to learning when your calculator gives an acceptable answer for a derivative and when it does not, you also need to understand the differences and similarities in mathematical derivative notation and calculator notation. The notation that is used for the calculator's numerical derivative is nDeriv($f(x)$, x, x). The correspondence between the mathematical derivative notation $\frac{df(x)}{dx}$ and the calculator's notation nDeriv($f(x)$, x, x) is illustrated in Figure 5.

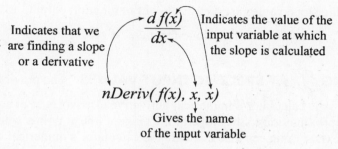

Figure 5

We illustrate another use of the TI-83's derivative by constructing Table 4.14 in Section 4.4 of *Calculus Concepts*. The table lists x, $y = f(x) = e^x$, and $y' = \frac{df}{dx}$ for seven different inputs.

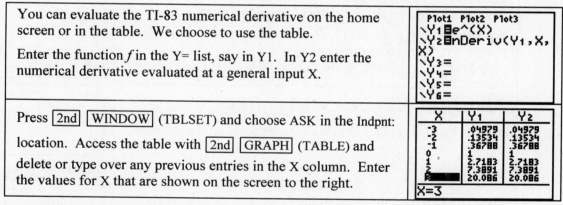

You can evaluate the TI-83 numerical derivative on the home screen or in the table. We choose to use the table.

Enter the function f in the Y= list, say in Y1. In Y2 enter the numerical derivative evaluated at a general input X.

Press [2nd] [WINDOW] (TBLSET) and choose ASK in the Indpnt: location. Access the table with [2nd] [GRAPH] (TABLE) and delete or type over any previous entries in the X column. Enter the values for X that are shown on the screen to the right.

It appears that the derivative values are the same as the function outputs. In fact, this is a true statement for all inputs of f – this function is its own derivative!

Press [Y=] and edit Y1 to be the function $g(x) = 2^x$.

Access the statistical lists, clear any previous entries from L1, L2, L3, and L4. Enter the *x*-values shown above in L1.

Highlight L2 and enter Y1(L1). Remember to type L1 using [2nd] 1 (L1).

Press ENTER to fill L2 with the function outputs. Then, highlight L3 and type Y2(L1).	
Press ENTER to fill L3 with the derivative of Y1 evaluated at the inputs in L1. Note that these are not the same as the function outputs. To see what relation the slopes have to the function outputs, press ▶ and highlight L4. Type L3 ÷ L2 ENTER.	

It appears that the slope values are a multiple of the function output. In fact, that multiple is $\ln 2 \approx 0.693147$. Thus we have this slope formula: If $g(x) = 2^x$, then $\frac{dg}{dx} = (\ln 2) 2^x$.

4.4.2 CALCULATING $\frac{dy}{dx}$ AT SPECIFIC INPUT VALUES

The previous two sections of this *Guide* examined the calculator's numerical derivative nDeriv($f(x)$, x, a) and illustrated that it gives a good approximation to the slope of the tangent line at points where the instantaneous rate of change exists. You can also evaluate the calculator's numerical derivative from the graphics screen using the CALC menu. However, instead of being named nDeriv(in that menu, it is called *dy/dx*. We illustrate this use with the function in part *a* of Example 2 in Section 4.4 of *Calculus Concepts*.

Enter $f(x) = 12.36 + 6.2 \ln x$ in Y1. (All previously-entered functions should be cleared except for this one in Y1.) We want to draw a graph of f. Realize that because of the log term, $x \geq 0$. Choose some value for Xmax, say 5. Then use ZOOM ▲ [ZoomFit] to set the height of the graph.	
With the graph on the screen, press 2nd TRACE (CALC) 6 [dy/dx]. Use ◀ or ▶ to move to some point on the graph.	
Press ENTER and the slope of the function is calculated at the input you chose in the previous step.	
If you want the derivative evaluated at a specific value of X, you could just type in the desired input instead of pressing the arrow keys. Press CLEAR and 2nd TRACE (CALC) 6 [dy/dx] 3 ENTER.	

> The slope $dy/dx = 2.0666667$ appears at the bottom of the screen. Return to the home screen and press $\boxed{X,T,\theta,n}$. The TI-83's X memory location has been updated to 3. Now type the numerical derivative instruction (evaluated at 3) as shown to the right. This is the dy/dx value you saw on the graphics screen.
>
> ```
> X            3
> nDeriv(Y₁,X,3)
>     2.066666743
> ```

You can use the ideas presented above to check your algebraic formula for the derivative. We next investigate this procedure.

4.4.3 NUMERICALLY CHECKING SLOPE FORMULAS
It is always a good idea to check your answer. Although your calculator cannot give you an algebraic formula for the derivative function, you can use numerical techniques to check your algebraic derivative formula. The basic idea of the checking process is that if you evaluate your derivative and the TI-83's numerical derivative at several randomly chosen values of the input variable and the outputs are basically the same values, your derivative is *probably* correct.

These same procedures are applicable when you check your results (in the next several sections) after applying the Sum Rule, the Chain Rule, or the Product Rule. We use the function in part *c* of Example 2 in Section 4.4 of *Calculus Concepts* to illustrate.

> Enter $m(r) = \dfrac{8}{r} - 12\sqrt{r}$ in Y1 (using X as the input variable).
>
> Compute $m'(r)$ using pencil and paper and the derivative rules. Enter this function in Y2. (What you enter in Y2 may or may not be the same as what appears to the right.)
>
> Enter the TI-83's numerical derivative of Y1 (evaluated at a general input X) in Y3. Because you are interested in seeing if the outputs of Y2 and Y3 are the same, turn off Y1.
>
> ```
> Plot1 Plot2 Plot3
> \Y1=8/X-12√(X)
> \Y2= -8/X²-6/√(X)
> \Y3=nDeriv(Y1,X,
> X)
> \Y4=
> \Y5=
> ```
>
> Press $\boxed{\text{2nd}}$ $\boxed{\text{WINDOW}}$ (TBLSET) and choose ASK in the Indpnt: location. Access the table with $\boxed{\text{2nd}}$ $\boxed{\text{GRAPH}}$ (TABLE) and delete or type over any previous entries in the X column. Enter at least three different values for X.
>
> | X | Y2 | Y3 |
> |---|---|---|
> | 1 | -14 | -14 |
> | 5.3 | -2.891 | -2.891 |
> | 12.86 | -1.722 | -1.722 |
> | 159 | -.4761 | -.4761 |
> | 2089 | -.1313 | -.1313 |
>
> X=

The table gives fairly strong evidence that that Y2 and Y3 are the same function.

4.4.4 GRAPHICALLY CHECKING SLOPE FORMULAS
When it is used correctly, a graphical check of your algebraic formula works well because you can look at many more inputs when drawing a graph than when viewing specific inputs in a table. We illustrate this use with the function in part *d* of Example 2 in Section 4.4 of *Calculus Concepts*.

> Enter $j(y) = 17{,}000\left(1 + \dfrac{0.025}{12}\right)^{12y}$ in Y1 (using X as the input variable). Compute $j'(y)$ using pencil and paper and the derivative rules. Enter this function in Y2.
>
> Enter the TI-83's numerical derivative of Y1 (evaluated at a general input X) in Y3. Before proceeding, turn off the graphs of Y1 and Y3.
>
> ```
> Plot1 Plot2 Plot3
> \Y1=17000(1+.025
> /12)^(12X)
> \Y2=17000*ln(1.0
> 25288)(1.025288^
> X)
> \Y3=nDeriv(Y1,X,
> X)
> ```

To graphically check your derivative formula answer, you now need to find a good graph of Y2. Because this function is not in a context with a given input interval, the time it takes to find a graph is shortened if you know the approximate shape of the graph. Note that the graph of the function in Y2 is an increasing exponential curve.

Start with ZOOM 4 [ZDecimal] or ZOOM 6 [ZStandard]. Neither of these shows a graph (because of the large coefficient in the function), but you can press TRACE to see some of the output values. Using those values, reset the window. The graph to the right was drawn in the window [−10, 10] by [330, 550], but any view that shows the graph will do.	
Now, press Y=, turn off Y2, and turn on Y3, which is the derivative computed by the TI-83. Press GRAPH. Note that you are drawing the graph of Y3 in *exactly the same window* in which you graphed Y2.	

If you see the same graph, your algebraic formula (in Y2) is very likely correct.

4.5 The Chain Rule

You probably noticed that checking your answer for a slope formula graphically is more difficult than checking your answer using the TI-83 table if you have to spend a lot of time finding a window in which to view the graph. Practice with these methods will help you determine which is the best to use to check your answer. (These ideas also apply to Section 4.6.)

4.5.1 SUMMARY OF CHECKING METHODS

Before you begin checking your answer, make sure that you have correctly entered the function. It is very frustrating to miss the answer to a problem because you have made an error in entering a function in your calculator. We summarize the methods of checking your algebraic answer using the function in Example 3 of Section 4.5 of *Calculus Concepts*.

Enter the function $P(t) = \dfrac{77.9}{1 + 39.5e^{-0.4984t}}$ in the Y1 location, your answer for P' in Y2, and the TI-83 derivative in Y3. Turn off Y1.	
Go to the table, which has been set to ASK mode. Enter at least 3 input values. (Because this problem is in a context, read the problem to see which inputs make sense.) It seems that the answer in Y2 is probably correct!	
If you prefer a graphical check, the problem states that the equation is valid between 1980 and 1996 (and the input is the number of years after 1980). So, turn off Y3, set Xmin = 0, Xmax = 16, and draw a graph of Y2 using ZoomFit. Then, using the same window, draw a graph of Y3 with Y2 turned off. The graphs are the same, again indicating a correct answer in Y2.	

Chapter 5   Analyzing Change: Applications of Derivatives

5.2 Relative and Absolute Extreme Points

Your calculator can be very helpful for checking your analytic work when you find optimal points and points of inflection. When you are not required to show work using derivative formulas or when an approximation to the exact answer is all that is required, it is a simple process to use your calculator to find optimal points and inflection points.

5.2.1 FINDING X-INTERCEPTS OF SLOPE GRAPHS   Where the graph of a function has a relative maximum or minimum, the slope graph has a horizontal tangent. Where the tangent line is horizontal, the derivative of the function is zero. Thus, finding where the slope graph *crosses* the input axis is the same as finding the input of a relative extreme point.

Consider, for example, the model for Acme Cable Company's revenue for the 26 weeks after it began a sales campaign, where x is the number of weeks since Acme began sales:

$$R(x) = -3x^4 + 160x^3 - 3000x^2 + 24{,}000x \text{ dollars}$$

In Example 2 of Section 5.2 of *Calculus Concepts*, we are asked to find when Acme's revenue was greatest during the 26-week interval.

Enter R in the Y1 location of the Y= list. Enter either the TI-83's derivative or your derivative in the Y2 location. Turn off Y1. (If you use your derivative, be sure to use one of the methods at the end of Chapter 4 in this *Guide* to check that your derivative and the TI-83 derivative are the same.)	
The statement of the problem indicates that x should be graphed between 1 and 26. Set this horizontal view, and draw the slope graph in Y2 with ZOOM ▲ [ZoomFit]. For a better view to use in this illustration, reset the window values to Ymin = -800 and Ymax = 3000. Redraw the slope graph with GRAPH.	
Find the intercepts of the slope graph using the CALC menu. With the graph on the screen, press 2nd TRACE (CALC) 2 [zero]. Use ▶ to move the cursor near to, but still to the *left* of, the rightmost x-intercept.	
Press ENTER to mark the location of the left bound. Use ▶ to move the cursor near to, but to the *right* of, the rightmost x-intercept.	

- Note that the calculator has marked (at the top of the screen) the left bound with a small triangle. The right bound will be similarly marked. The x-intercept must lie between these two bound marks. If you incorrectly mark the interval, you may not get an answer.

Press [ENTER] to mark the location of the right bound. You are next asked to provide a guess. Any value near the intercept will do. Use [◄] to move the cursor near the intercept.

Press [ENTER]. The location of the x-intercept is displayed.

We see that $R'(x) = 0$ at $x = 20$.

(The numerical process used the find the zero sometimes includes small rounding errors. The output (y) at the root should be 0 but is printed on the screen as -0.00003.)

We now need to determine if the slope graph crosses the x-axis, only touches the x-axis, or does neither of these at the other location that may be an intercept.

Use ZoomIn (see Section 1.4.2) or ZBox (see Section 3.2.1) as many times as necessary to magnify the portion of the graph around $x = 20$ in order to examine it more closely.

We choose to use ZBox, but both work equally well.

After magnifying the graph several times, we see that the graph just touches and does not cross the x-axis near $x = 10$.

(The coordinates at the bottom of the screen give the location of the cursor, which may or may not be on the graph.) We see that $x = 10$ does not yield an extreme point on the graph of R.

We are asked to find the absolute maximum, and we know that it occurs at one of the endpoints of the interval or at a zero of the slope graph. So, return to the home screen and find the outputs of R at the endpoints of the interval ($x = 0$ and $x = 26$) as well as the output at the "crossing" zero of the slope graph ($x = 20$).

We see that Acme's revenue was greatest at 20 weeks after they began the sales campaign.

5.2.2 FINDING ZEROS OF SLOPE FUNCTIONS USING THE SOLVER
You may find it more convenient to use the TI-83 SOLVER rather than the graph and the CALC menu to find the x-intercept(s) of the slope graph. We illustrate using Acme Cable's revenue function.

Enter R in the Y1 location of the Y= list. Enter either the TI-83's derivative or your derivative in the Y2 location.

Turn off Y2.

Draw a graph of Y1 to obtain a guess as to the location of any relative maxima for use in the SOLVER. The statement of the problem indicates that x should be graphed between 1 and 26. Set this horizontal view, and draw the function graph with [ZOOM] [▲] [ZoomFit] [ENTER].

TI-83 Guide

A-61

Reset Ymax to a larger value, say 95,000, to better see the high point on the graph. Graph R and press TRACE. Hold down ▶ until you have an estimate of the input location of the high point. The maximum seems to occur when x is near 19.9.	Y1=-3X^4+160X^3-3000X2+_ X=19.882979 Y=79991.911
Access the SOLVER with MATH 0 and press ▲ to go to the eqn: location. Clear what is there from a previous problem and enter Y2. (Recall that we want to solve the equation $R'(x) = 0$ to find where the tangent line is horizontal.)	EQUATION SOLVER eqn:0=Y2
Press ENTER. Type in the guess that you obtained for the location of the maximum by tracing the graph of R. With the cursor on the line containing X, press ALPHA ENTER (SOLVE).	Y2=0 X=19.9■ bound={-1E99,1…
The solution $x = 20$ is found. Recall that calculators use numerical algorithms to find zeros. You may or may not obtain the exact value 20. Always round the answer obtained from the SOLVER to make sense in the problem context (here, round to a whole number.)	Y2=0 •X=19.999999969… bound={-1E99,1… •left-rt=1E-5
Note that if you enter $x \approx 10$ as a guess in the SOLVER, the SOLVER will still return the solution $x = 20$. Acme's revenue was greatest at 20 weeks after they began the sales campaign.	Y2=0 •X=19.999999939… bound={-1E99,1… •left-rt=-2E-5

5.2.3 USING THE CALCULATOR TO FIND OPTIMAL POINTS
Once you draw a graph of a function that clearly shows the optimal points, your calculator can find the location of those high points and low points without using calculus. However, we recommend not relying only on this method because your instructor may ask you to show your work using derivatives. If so, this method would probably earn you no credit! This method does give a good check of your answer, and we illustrate it using Acme Cable's revenue function R from Section 5.2.1.

Enter R in the Y1 location of the Y= list. The statement of the problem indicates that x should be graphed between 1 and 26. Set this horizontal view, and draw the function graph with ZOOM ▲ [ZoomFit] ENTER. Press WINDOW and reset Ymax to 95,000. Press GRAPH.	
With the graph of R on the screen, press 2nd TRACE (CALC) 4 [maximum]. Use ▶ to move the cursor near, but still to the *left* of, the high point on the curve.	Y1=-3X^4+160X^3-3000X2+_ LeftBound? X=17.223404 Y=76908.491

Copyright © Houghton Mifflin Company. All rights reserved.

Press ENTER to mark the left bound of the interval. Use ▶ to move the cursor to the *right* of the high point on the curve. Press ENTER to mark the right bound of the interval. Use ◀ to move the cursor to your guess for the high point on the curve.	

CAUTION: Notice the small arrowheads at the top of the screen that mark the bounds of the interval. The TI-83 returns the highest point that is within the bounded interval that you have marked when you press ENTER. If the relative maximum does not lie in this interval, the TI-83 will return, instead, the highest point in the interval you marked (usually an endpoint).

Press ENTER and the location of the relative maximum (x) and the relative maximum value (y) are displayed. (As previously mentioned, you may not see the exact answer due to rounding errors in the numerical routine used by the TI-83. Always round your answer to make sense in the context.)	

- The method shown in this section also applies to finding the relative minimum values of a function. The only difference is that to find the relative minimum instead of the relative maximum, initially press 2nd TRACE (CALC) 3 [minimum].

5.4 Inflection Points

As was the case with optimal points, your calculator can be very helpful in checking algebraic work when you find points of inflection. You can also use the methods illustrated in Section 5.2.3 of this *Guide* to find the location of any maximum or minimum points on the graph of the first derivative to find the location of any inflection points for the function.

5.4.1 FINDING X-INTERCEPTS OF A SECOND-DERIVATIVE GRAPH We first look at using the algebraic method of finding inflection points – finding where the graph of the second derivative of a function *crosses* the input axis.

To illustrate, we consider a model for the percentage of students graduating from high school in South Carolina from 1982 through 1990 who entered post-secondary institutions:

$$f(x) = -0.1057x^3 + 1.355x^2 - 3.672x + 50.792 \text{ percent}$$

where x is the number of years after 1982.

Enter f in the Y1 location of the Y= list, the first derivative of f in Y2, and the second derivative of f in Y3. (Be careful not to round any decimal values.) Turn off Y1 and Y2.	

We are given the input interval 1982 through 1990, so $0 \leq x \leq 8$. Either use ZoomFit or choose some appropriate vertical view. We use $-4 \leq y \leq 4$. Because we are looking for the x-intercept(s) of the second derivative graph, any view that shows the line crossing the horizontal axis is okay to use.	
Use the methods indicated in Section 5.2.1 to find where the second derivative graph crosses the x-axis. (Note that when you are asked for the inflection *point* of f, give both the input and an output of the original function.)	
Return to the home screen and enter $\boxed{X,T,\theta,n}$. The x-value you just found as the x-intercept remains stored in the X location until you change it by tracing, using the SOLVER, and so forth. Find the y-value by substituting this x-value into Y1.	
At some point, be sure to examine a graph of the function and verify that an inflection point does occur at the point you have found. To do this, turn off Y3, turn on Y1, and use ZoomFit to draw the graph. Trace near where $x \approx 4.27$ and $y \approx 51.6$. The graph of R confirms that an inflection point occurs at this point.	

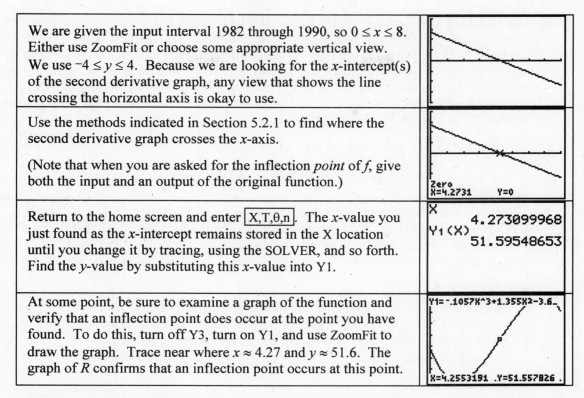

EXPLORE: What would the line tangent to the graph of f look like at the inflection point? Use the graph of Y1 and the first method explained on page A-49 of this *Guide* (with $x = 4.27$) to see if you are correct.

EXPLORE: What is true about the graph of f' at the inflection point? Use the graph of Y2 and the trace cursor to see if you are correct.

- The TI-83 will usually draw an accurate graph of the first derivative of a function when you use nDeriv(. However, this calculator does not have a built-in method to calculate or graph f'', the second derivative. As illustrated below, you can try to use nDeriv(f', X, X) to find f''. Be cautioned, however, that nDeriv(f', X, X) sometimes "breaks down" and gives invalid results. If this should occur, the graph of nDeriv(f', X, X) appears very jagged and this method should not be used.

Enter f in the Y1 location of the Y= list, the first derivative of f in Y2, and the second derivative of f in Y3, using the TI-83's numerical derivative for each derivative that you enter. Turn off Y1. Be sure that Xmin = 0 and Xmax = 8 (as the problem directions indicate.)	
Draw the graph of the first derivative (Y2) and the second derivative (Y3) of f using an appropriate window. You can use ZoomFit to set the vertical view or experiment until you find a suitable view. The graph to the right uses $-4 \leq y \leq 4$.	

Find the x-intercept of the second derivative graph as indicated in this section or find the input of the high point on the first derivative graph (see Section 5.2.3) to locate the inflection point.

5.4.2 USING THE CALCULATOR TO FIND INFLECTION POINTS

Remember that an inflection point on the graph of a function is a point of greatest or least slope. Whenever finding the second derivative of a function is tedious algebraically and/or you do not need an exact answer from an algebraic solution, you can easily find the input location of an inflection point by finding where the first derivative of the function has a maximum or minimum slope.

We illustrate this method using the logistic function for polio cases that is in Example 2 of Section 5.4 in *Calculus Concepts*:

The number of polio cases in the U.S. in 1949 is given by $C(t) = \dfrac{42{,}183.911}{1+21{,}484.253e^{-1.248911t}}$

where $t = 1$ in January, $t = 2$ in February, and so forth.

Enter C in the Y1 location of the Y= list and the first derivative of f in Y2. (You can use your algebraic formula for the first derivative or the calculator's numerical derivative.) Turn off Y1.
The problem context says that the input interval is from 0 (the beginning of 1949) to 12 (the end of 1949), so set these values for x in the WINDOW. Set the vertical view and draw the graph of C' with ZoomFit.
Use the methods discussed in Section 5.2.3 ($\boxed{\text{2nd}}$ $\boxed{\text{TRACE}}$ (CALC) 4 [maximum]) to find the input location of the maximum point on the slope graph.
The x-value of the maximum of the slope graph is the x-value of the inflection point of the function. To find the rate of change of polio cases at this time, substitute this value of x in Y2. To find the number of cases at this time, substitute x in Y1.

CAUTION: Do not forget to round your answers appropriately (this function should be interpreted discretely) and to give units of measure with each answer.

Note that you could have found the input of the inflection point on the polio cases graph by finding the x-intercept of the second derivative graph. The function that is graphed to the right is Y3 = nDeriv(Y2, X, X), and the graph was drawn using ZoomFit.
If you prefer, you could have found the input of the inflection point by solving the equation $C'' = 0$ using the SOLVER. (Do not forget that drawing the graph of C and tracing it can be used to find a guess for the SOLVER.)

Chapter 6   Accumulating Change: Limits of Sums and the Definite Integral

6.1 Results of Change and Area Approximations

So far, we have used the TI-83 to investigate rates of change. In this chapter we consider the second main topic in calculus – the accumulation of change. You calculator has many useful features that will assist in your investigations of the results of change.

6.1.1 AREA APPROXIMATIONS USING LEFT RECTANGLES
The TI-83 lists can be used to approximate, using left rectangles, the area between the horizontal axis and a rate of change function between two specified input values. We illustrate the necessary steps using the data for the number of customers entering a large department store during a Saturday sale. These data appear in Table 6.1 of Section 6.1 in *Calculus Concepts*.

Minutes after 9 a.m.	0	45	75	120	165	195	255	330	370	420	495	570	630	675
Customers per minute	1	2	3	4	4	5	5	5	5	4	4	3	2	2

Enter these data in lists L1 and L2. Find a cubic function to fit the data, and paste the function in Y1. Turn on Plot1. (If you did not obtain the function shown to the right, check your data values.) Next we graph the function on a scatter plot of the data over the 12-hour (720 minute) sale.	CubicReg y=ax³+bx²+cx+d a=4.5890356E -8 b=-7.781267E -5 c=.0330326126 d=.8876300878
Either set Xmin = 0, Xmax = 720, Xscl = 60 and use [ZOOM] [▲] [ZoomFit] [ENTER] to set the vertical view or press [ZOOM] 9 [ZoomStat] [ENTER]. Set Ymin = 0 and press [GRAPH].	

We want to approximate the area under the cubic graph and above the *x*-axis during the 12-hour time period using rectangles of equal width (60 minutes).

Press [STAT] 1 [EDIT], clear lists L1 and L2, and enter the inputs 0, 60, 120, …, 720 in L1 A quick way to do this is to highlight L1 and press [2nd] [STAT] (LIST) [▶] [OPS] 5 [seq(] [X,T,θ,n] [,] [X,T,θ,n] [,] 0 [,] 720 [,] 60 [)]. (Refer to Section 2.1.1 for more explanation on using the sequence command.)	L1 L2 L3 1 0 60 120 180 240 300 360 L1 ={0,60,120,18…
Because we are using rectangles with heights determined at the left endpoint of each interval, delete the last value in L1 (720). Substitute each input in L1 into the function in Y1 by highlighting L2 and typing Y1(L1). Press [ENTER].	L1 L2 L3 2 360 4.8359 420 4.4351 480 3.8904 540 3.2611 600 2.607 660 1.9873 ------ ------ L2(12)=1.98725049…

Consider what is now in the lists. L1 contains the left endpoints of the 12 rectangles and L2 contains the heights of the rectangles. If we multiply the heights by the widths of the rectangles (60 minutes) and enter this product in L3, we will have the rectangle areas in L3.

Copyright © Houghton Mifflin Company. All rights reserved.

Highlight L3 and type 60 [X] [2nd] 2 (L2) [ENTER]. L3 now contains the areas of the 12 rectangles.	L1 L2 L3 3 0 .88763 53.258 60 2.5994 155.96 120 3.8103 228.62 180 4.58 274.8 240 4.9678 298.07 300 5.0333 302 360 4.8359 290.15 L3(1)=53.25780526...
All that remains is to add the areas of the 12 rectangles. To do this, return to the home screen and type [2nd] [STAT] (LIST) [►] [►] [MATH] 5 [sum] [2nd] 3 (L3) [)] [ENTER]. We estimate that 2574 customers came to the Saturday sale.	sum(L3) 2573.712319

6.1.2 AREA APPROXIMATIONS USING RIGHT RECTANGLES
When you use left rectangles to approximate the results of change, the rightmost data point is not the height of a rectangle and is not used in the computation of the left-rectangle area. Similarly, when using right rectangles to approximate the results of change, the leftmost data point is not the height of a rectangle and is not used in the computation of the right-rectangle area. We illustrate the right-rectangle approximation with the data in Table 6.3 in Example 2 of Section 6.1.

End of day	1	5	9	13	17	21	25	29
Concentration rate of change (μg/mL/day)	1.5	0.75	0.33	0.20	0.10	−1.1	−0.60	−0.15

The data show the rate of change of the concentration of a drug in the blood stream in terms of the number of days since the drug was administered. Part *a* of Example 2 says to use the data and right rectangles of width 4 days to estimate the change in concentration of the drug from the end of day 1 through the end of day 29. Enter these data in lists L1 and L2.

Because the leftmost data point is not used with a right rectangle approximation, delete (1, 1.5) from lists L1 and L2. List L2 now contains the (signed) heights of the right rectangles. Highlight L3 and press 4 [2nd] 2 (L2) [ENTER]. List L3 now contains the (signed) rectangle areas.	L1 L2 L3 3 5 .75 3 9 .33 1.32 13 .2 .8 17 .1 .4 21 −1.1 −4.4 25 −.6 −2.4 29 −.15 −.6 L3 =4L2
Return to the home screen and find the sum of the seven right-rectangle areas with [2nd] [STAT] (LIST) [►] [►] [MATH] 5 [sum] [2nd] 3 (L3) [)] [ENTER]. The change in concentration was about −1.88 μg/mL.	sum(L3) −1.88
Part *b* of Example 2 says to find a model for the data. Return to the lists. Insert the first data point (1, 1.5) by placing the cursor over the first entries in L1 and L2 and pressing [2nd] [DEL] (INS). Type the data value over the 0 that is inserted in each. A scatter plot shows that a piecewise model with a break point at $x = 20$ is appropriate.	

Delete the last three data points from lists L1 and L2.

TI-83 Guide

Fit an exponential model to the data for days 1 through 17 (the five remaining data values in L1, L2), and put the function in Y1.

Type in the data for days 21, 25, and 29 in lists L3 and L4. Find a linear function for these data and paste the equation in Y2 with [STAT] ▶ 4 [LinReg(ax+b)] [2nd] 3 (L3) [,] [2nd] 4 (L4) [,] [VARS] ▶ [Y–VARS] 1 [Function] 2 [Y2].

```
Plot1 Plot2 Plot3
\Y1=1.7081947918
826*.84497654783
236^X
\Y2=.11875X+-3.5
854166666667
\Y3=
\Y4=
```

Refer to Section 1.3.8 of this *Guide* if you want to review the instructions for graphing this piecewise continuous function on a scatter plot of the data.

Part *c* of Example 2 says to find the change in the drug concentration from $x = 0$ through $x = 20$ using right rectangles of width 2 days. Enter the right endpoints (2, 4, 6, ..., 20) in L5.

Use the exponential function in Y1 to find the rectangle heights. Enter the heights in L6.

```
L4      L5      L6    6
-1.1    2       1.2196
-.6     4       .8708
-.15    6       .62174
------  8       .44391
        10      .31695
        12      .22629
        14      .16157
L6 =Y1(L5)
```

Find the right-rectangle areas by multiplying each entry in L6 by 2. This is the same as multiplying the sum of the heights by 2. On the home screen, press 2 [×] [2nd] [STAT] (LIST) ▶ ▶ [MATH] 5 [sum] [2nd] 6 (L6) [)] [ENTER].

```
2*sum(L6)
      8.234813648
```

6.1.3 AREA APPROXIMATIONS USING MIDPOINT RECTANGLES

Areas of midpoint rectangles are found using the same procedures as those used to find left and right rectangle areas except that the midpoint of the base of each rectangle is in L1 and no data values are deleted. We illustrate the midpoint-rectangle approximation using the data in Table 6.6 in Example 3 of Section 6.1. First, we align the input of Table 6.6 as shown below.

Number of hours since 11:45 a.m. Wednesday	0	4	8	12	16	20
Flow rate (cfs)	785	753	834	1070	1470	2090

Enter the data in the above table into lists L1 and L2. Because area units are height units times width units, we need to convert the output data in L2 to cubic feet per hour by multiplying each value by 3600 (the number of seconds in one hour).

```
L1      L2      L3    2
0       785     ------
4       753
8       834
12      1070
16      1470
20      2090
------
L2 =3600*L2
```

Press [ENTER]. After conversion, the units for the area are $\frac{\text{cubic feet}}{\text{second}} \cdot \frac{3600 \text{ seconds}}{1 \text{ hour}} \cdot \text{hours} \Rightarrow \text{cubic feet}$. A scatter plot of the converted data shows that a quadratic function is appropriate.

Find the quadratic equation and paste it into Y1.

```
Plot1 Plot2 Plot3
\Y1=18225X^2+-13
5334.28571429X+2
881542.8571429
\Y2=
\Y3=
\Y4=
\Y5=
```

To use 10 intervals and midpoint rectangles to approximate the area under the graph of the equation from $x = 0$ through $x = 20$, the width of each rectangle is 2 hours. Enter the midpoints 1, 3, 5, ..., 19 in L3. The heights of the rectangles are determined by the function in Y1, so evaluate Y1 at each of the values in L3 to find the rectangle heights.	L2 L3 L4 4 2.83E6 1 2.76E6 2.71E6 3 2.64E6 3E6 5 2.66E6 3.85E6 7 2.83E6 5.29E6 9 3.14E6 7.52E6 11 3.6E6 ------- 13 4.2E6 L4 =Y1(L3)

You can multiply each rectangle height by the width (2 hours), enter these values in another list, and sum the rectangle areas or you can multiply the sum of the heights in L4 by 2. Either calculation results in the area estimate. (To access the sum instruction, press [2nd] [STAT] (LIST) [▶] [▶] [MATH] 5 [sum].)	sum(L4)
	          39521250
	2*Ans
	          79042500

6.1.4 SIMPLIFYING AREA APPROXIMATIONS

The procedures used in the previous three sections of this *Guide* can become tedious when the number of rectangles is large. When you have a function $y = f(x)$ in the Y1 location of the Y= list, you will find program NUMINTGL very helpful in determining left-rectangle, right-rectangle, and midpoint-rectangle approximations for accumulated change. Program NUMINTGL (listed in the *TI-83 Program Appendix*) performs automatically all the calculations that you have been doing manually using the lists.

WARNING: This program will not execute properly unless the function that determines the heights uses X as the input variable and is pasted in Y1. If you receive an error message while attempting to run this program, consult Programs in *TroubleShooting the TI-83* in this *Guide*.

We illustrate using this program with the function for the Carson River flow rates that was fit to data in the previous section.

$$f(h) = 18{,}225h^2 - 135{,}334.3h + 2{,}881{,}542.9 \text{ cubic feet per hour}$$

h hours after 11:45 a.m. Wednesday. (The complete function found from the data in Table 6.6 is used for all following calculations. It should be in the Y1 location of the Y= list.)

Press [PRGM] followed by the number or letter next to the location of the program and [ENTER] to start NUMINTGL. (At this point, if you did not enter the function in Y1, press 2 [ENTER] to exit the program. Enter the function in Y1 and re-run the program). If the function is in Y1, press 1 [ENTER] to continue.	ENTER F(X) IN Y1 CONTINUE? YES(1) NO(2) ■
At the next prompt, press 1 [ENTER] to draw the approximating rectangles or press 2 [ENTER] to obtain only numerical approximations to the area between the function and the horizontal axis between 0 and 20. For illustration purposes, let's choose to see the rectangles.	ENTER F(X) IN Y1 CONTINUE? YES(1) NO(2) 1 DRAW PICTURES? YES(1) NO(2) 1
At the LEFT ENDPOINT? prompt, press 0 [ENTER], and at the RIGHT ENDPOINT? prompt, press 20 [ENTER] to tell the TI-83 the input interval. You are next shown a menu of choices. Press 4 and [ENTER] for a midpoint-rectangle approximation.	ENTER CHOICE: LEFT RECT (1) RIGHT RECT (2) TRAPEZOIDS (3) MIDPT RECT (4) 4

Input 10 at the N? prompt and press ENTER. A graph of the 10 approximating midpoint rectangles and the function are shown. (Note that the program automatically sets the height of the window based on the left and right endpoints of the input interval.)	
The midpoint-rectangle area is displayed as SUM when you press ENTER.	
Press ENTER once more and some choices are displayed. Suppose you now want to find the approximating area and see the figure for 10 right rectangles. Press 2 to change the method, and then press 2 ENTER to choose right rectangles.	
Press 10 ENTER at the N? prompt to view the right rectangles.	
Again press ENTER and the right-rectangle area estimate is displayed.	
Continue in this manner and find the left-rectangle area or change N and find the left-, right-, or midpoint-rectangle approximations for different numbers of subintervals. When you finish, press ENTER and choose 3 to QUIT the program.	

6.2 The Definite Integral as a Limit of Sums

This section introduces you to a very important and useful concept of calculus -- the definite integral. Your calculator can be very helpful as you study definite integrals and how they relate to the accumulation of change.

6.2.1 LIMIT OF SUMS When you are looking for a trend in midpoint-rectangle approximations to the area (or signed area) between a function and the horizontal axis between two values of the input variable, program NUMINTGL is extremely useful! However, when finding a limit of sums using this program, it is not advisable to draw pictures when N, the number of rectangles, is large unless you have a lot of time to wait for many, many rectangles to be drawn.

We illustrate using this program to find a limit of sums using $f(x) = \sqrt{1-x^2}$, the function in Example 1 of Section 6.2 of *Calculus Concepts*.

To construct a table of midpoint-rectangle approximations for the area between f and the x-axis from $x = 0$ to $x = 1$, first clear the Y= list, enter f in Y1, and turn off Plot 1. (Type the square root symbol with [2nd] [x^2].)	Plot1 Plot2 Plot3 \Y1■√(1-X²) \Y2= \Y3= \Y4= \Y5= \Y6= \Y7=
Run program NUMINTGL. At the first prompt, enter 1 to continue and enter 2 to not draw pictures.	ENTER F(X) IN Y1 CONTINUE? YES(1) NO(2) 1 DRAW PICTURES? YES(1) NO(2) 2
At the LEFT ENDPOINT? prompt, enter 0, and at the RIGHT ENDPOINT? prompt, enter 1 to tell the TI-83 that the input interval is from $x = 0$ to $x = 1$. You are next shown a menu of choices. Always use midpoint rectangles with a limit of sums. Press 4 and [ENTER]. At the N? prompt, enter 4.	ENTER CHOICE: LEFT RECT (1) RIGHT RECT (2) TRAPEZOIDS (3) MIDPT RECT (4) 4 N? 4■
Record on paper the area estimate for 4 midpoint rectangles. (You should record at least one more decimal place than the number of places needed for the required accuracy. Because Example 1 asks for three decimal-place accuracy, record at least four decimal places of this value.)	RIGHT RECT (2) TRAPEZOIDS (3) MIDPT RECT (4) 4 N? 4 SUM= .7959823052
Press [ENTER] to continue the program, and press 1 to choose CHANGE N. Now double the number of rectangles by entering N = 8. Record the area approximation (again, to four decimal places).	N? 8 SUM= .7891717328
Continue in this manner, each time choosing the first option, CHANGE N, and doubling N until a trend is evident. *Note:* Intuitively, finding a *trend* means that you are sure what the area approximation, to three decimal places (or whatever accuracy is specified in the problem), will be without having to use larger values of N in the program.	N? 256 SUM= .7854191806

6.4 The Fundamental Theorem

The Fundamental Theorem of Calculus is important because it connects the two main topics in calculus – differentiation and integration. In this section of the *Guide*, we see how to use the TI-83's definite integral function and use it along with the TI-83's numerical derivative to illustrate the Fundamental Theorem in action.

6.4.1 DEFINITE INTEGRAL NOTATION AND CALCULATOR NOTATION

Recall that nDeriv is the TI-83's numerical derivative and provides, in most cases, a very good approximation for the instantaneous rate of change of a function when that rate of change exists. As is the case with the numerical derivative, your calculator does not give a formula for accumulation functions. However, it does give an excellent numerical estimate for the definite integral of a function between two specific input values when that integral exists.

The TI's numerical integral is called fnInt, and the correspondence between the mathematical notation $\int_a^b f(x)\,dx$ and the calculator's notation fnInt(f(x), x, a, b) is illustrated in Figure 6.

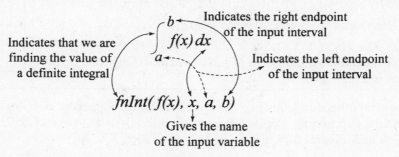

Figure 6

We illustrate the use of the calculator's definite integral function by finding the area from $x = 0$ to $x = 1$ between the function $f(x) = \sqrt{1-x^2}$ and the horizontal axis. Recall that the trend (to three decimal places) found from the limit of sums investigation in Example 1 of Section 6.2 of *Calculus Concepts* was 0.785. (Also see Section 6.5.1 of this *Guide*.)

You can enter f in Y1, or any of the other function locations, and refer to it in the fnInt expression as Y1 or you can enter the function directly on the home screen as shown to the right.
 	
 Access fnInt with MATH 9 [fnInt(].
 	
 Note that when you type the function on the home screen, you can use any letter as the input variable. In the Y= list, you must use X as the variable name.
	`fnInt(√(1-X²),X,`
	`0,1)`
	`        .7853984608`
	`fnInt(√(1-T²),T,`
	`0,1)`
	`        .7853984608`

6.4.2 THE FUNDAMENTAL THEOREM OF CALCULUS Intuitively, this theorem tells us that the derivative of an antiderivative of a function is the function itself. Let us view this idea both numerically and graphically. The correct syntax for the TI-83's numerical integrator is

fnInt(function, name of input variable, left endpoint for input, right endpoint for input)

Consider the function $f(t) = 3t^2 + 2t - 5$ and the accumulation function $F(x) = \int_1^x f(t)\,dt$.

The Fundamental Theorem of Calculus tells us that $F'(x) = \dfrac{d}{dx}\left(\int_1^x f(t)\,dt\right) = f(x)$; that is, $F'(x)$ is f evaluated at x.

Input f in Y1 and F' in Y2 (remember that the TI-83 requires that you use X as the input variable in the Y= list).
 	
 Access fnInt with MATH 9 [fnInt(] and nDeriv with MATH 8 [nDeriv(]. (Turn off any stat plots that are on.)
	`Plot1 Plot2 Plot3`
	`\Y1■3X²+2X-5`
	`\Y2■nDeriv(fnInt`
	`(Y1,X,1,X),X,X)`
	`\Y3=`
	`\Y4=`
	`\Y5=`
	`\Y6=`

Have TBLSET set to ASK and press 2nd GRAPH (TABLE). Input several different values for X. Other than occasional roundoff error because the calculator is approximating these values, the results are identical.	X / Y1 / Y2 -5 / 60 / 60 0 / -5 / -5 4.3 / 59.07 / 59.07 159.82 / ..... / 76942 1.6667 / 6.6667 / 6.6667 `Y1=76941.9372`

Find a suitable viewing window such as the one set with ZOOM 6 [ZStandard]. Without changing the window (that is, draw the graphs by pressing GRAPH), turn off Y2 and draw the graph of Y1. Then turn off Y1 and draw the graph of Y2.

(*Note:* The graph of Y2 takes a while to draw.) Turn both Y1 and Y2 on and draw the graph of both functions. Only one graph is seen in each case.

Enter several other functions in Y1 and do not change Y2 except possibly for the left endpoint 1 in the fnInt expression. Perform the same explorations as above. Confirm your results with derivative and integral formulas.

6.4.3 DRAWING ANTIDERIVATIVE GRAPHS

Recall when using fnInt($f(x)$, x, a, b) that a and b are, respectively, the lower and upper endpoints of the input interval. Also remember that you do not have to use x as the input variable unless you are graphing the integral or evaluating it using the calculator's table. Unlike when graphing using nDeriv, the TI-83 will not graph a general antiderivative; it only draws the graph of a specific accumulation function. Thus, we can use x for the input at the upper endpoint when we want to draw an antiderivative graph, but not for the inputs at both the upper and lower endpoints.

All of the antiderivatives of a specific function differ only by a constant. We explore this idea using the function $f(x) = 3x^2 - 1$ and its general antiderivative $F(x) = x^3 - x + C$. Because we are working with a general antiderivative in this illustration, we do not have a starting point for the accumulation. We therefore choose some value, say 0, to use as the starting point for the accumulation function to illustrate drawing antiderivative graphs. If you choose a different lower limit, your results will differ from those shown below by a constant.

Enter f in Y1, fnInt(Y1, X, 0, X) in Y2, and F in Y3, Y4, Y5, and Y6, using a different number for C in each function location. (You can use the values shown to the right or different values.)	Plot1 Plot2 Plot3 \Y1■3X²-1 \Y2■fnInt(Y1,X,0,X) \Y3■X^3-X+1 \Y4■X^3-X-2 \Y5■X^3-X+5 \Y6■X^3-X-4.8
Find a suitable viewing window and graph the functions Y1 through Y6. The graph to the right was drawn with $-3 \leq x \leq 3$ and $-20 \leq y \leq 20$. It appears that the only difference in the graphs of Y2 through Y6 is the *y*-axis intercept. But, isn't C the *y*-axis intercept of each of these antiderivative graphs?	
Clear Y4, Y5, and Y6. Turn off Y1 and change the 1 in Y3 to 0.	Plot1 Plot2 Plot3 \Y1=3X²-1 \Y2■fnInt(Y1,X,0,X) \Y3■X^3-X+0 \Y4= \Y5= \Y6=
Press GRAPH and draw the graphs of Y2 and Y3. You should see only one graph. Set the TI-83 TABLE to ASK and enter some values for x. It appears that Y2 and Y3 are the same function.	X | Y2 | Y3 -5 | -120 | -120 0 | 0 | 0 4.3 | 75.207 | 75.207 -8.35 | -573.8 | -573.8 21 | 9240 | 9240 51.2 | 134167 | 134167 X=

CAUTION: The methods for checking derivative formulas that were discussed in Sections 4.4.3 and 4.4.4 are not valid for checking general antiderivative formulas. Why not? Because to graph an antiderivative using fnInt, you must arbitrarily choose values for the constant of integration and for the input of the lower endpoint. However, for most of the rate-of-change functions where $f(0) = 0$, the calculator's numerical integrator values and your antiderivative formula values should differ by the same constant at every input value where they are defined.

6.5 The Definite Integral

When using the numerical integrator on the home screen, enter fnInt($f(x)$, x, a, b) for a specific function f with input x and specific values of a and b. (Remember that the input variable does not have to be x when on the home screen.) If you prefer, f can be in the Y= list and referred to as Y1 (or whatever location the function is in) when using fnInt.

6.5.1 EVALUATING A DEFINITE INTEGRAL ON THE HOME SCREEN We illustrate the use of fnInt with the function that models the rate of change of the average sea level. The rate-of-change data are given in Table 6.43 of Example 3 in Section 6.5 of *Calculus Concepts*.

Time (thousands of years before the present)	-7	-6	-5	-4	-3	-2	-1
Rate of change of average sea level (meters/year)	3.8	2.6	1.0	0.1	-0.6	-0.9	-1.0

Enter the time values in L1 and the rate of change of the average sea level values in L2.

A scatter plot of the data indicates a quadratic function. Find the function and paste it in Y1. (Draw the function on the scatter plot of the data to confirm that it gives a good fit.)	
Because part *b* of Example 3 asks for the *areas* of the regions above and below the input axis and the function, we must find where the function crosses the axis. You can find this value using the solver (solve Y1 = 0) or by using the graph and the *x*-intercept method described in Section 1.1.9 of this *Guide*.	

CAUTION: If you round the *x*-intercept value, it will cause whatever you do with this value to not result in as accurate a value as it could be. This situation occurs many times in this and the next several chapters of the text. Also recall that one of the numerical considerations given in Chapter 1 of *Calculus Concepts* is that intermediate calculation values should not be rounded. For maximum accuracy, store this value in some memory location, say Z, and refer to it as Z in all subsequent calculations.

Whether you use the SOLVER or the *x*-intercept method shown on the calculator screen above, the zero is stored in X. Return to the home screen and store the value in Z (or whatever memory location, except X, that you choose). Do not use the X location because the X value changes whenever you use the TI-83 table, the solver, or trace a graph. Type Z with ALPHA 2.	X→Z -3.844955338

Find the area of the region above the horizontal axis by typing and entering the first expression shown on the right.
 	
Find the area of the region below the horizontal axis by typing and entering the second expression shown on the right. (The negative is used because the region is below the input axis.)
	fnInt(Y₁,X,-7,Z)
	        5.405934876
	-fnInt(Y₁,X,Z,0)
	        2.936490432

Part *c* of Example 3 asks you to evaluate $\int_{-7}^{0} Y1\, dx$. Find this value by typing and entering the expression shown on the right. Note that the result is not the sum of the two areas – it is their difference.	fnInt(Y₁,X,-7,0) 2.469444444

- If you evaluate a definite integral using antiderivative formulas and check your answer with the calculator using fnInt, you might find a slight difference in the last few decimal places. Remember that the TI-83 is evaluating the definite integral using an approximation technique, not an algebraic formula.

6.5.2 EVALUATING A DEFINITE INTEGRAL FROM THE GRAPHICS SCREEN

The value of a definite integral can also be found from the graphics screen. We again illustrate the use of fnInt with the function that models the rate of change of the average sea level.

Turn off Plot 1 and have the function modeling the average sea level in Y1. (See Section 6.5.1 of this *Guide*.) We want to evaluate $\int_{-7}^{0} Y1\, dx$. (You should practice using this method by finding the areas of the regions indicated in part *b* of Example 3 in Section 6.5 of the text.)
Set Xmin = -7, Xmax = 0, and draw the graph of Y1 with ZOOM ▲ [ZoomFit] ENTER. Press 2nd TRACE (CALC) 7 [∫f(x)dx]. The TI-83 asks Lower Limit? Press (-) 7 and obtain the screen shown to the right.
Press ENTER. The TI-83 now asks Upper Limit? Press 0 and obtain the screen shown to the right.
Press ENTER. The region between the function and the horizontal axis from the lower limit to the upper limit is shaded. The value of the integral is shown at the bottom of the screen.

CAUTION: If you type in a value for the lower and/or upper limit (that is, the input at the endpoint) that is not visible on the graphics screen, you will get an error message when you attempt to evaluate the integral. Be certain that these limits are included in the interval from Xmin to Xmax before using this method.

Chapter 7 Analyzing Accumulated Change: Integrals in Action

7.1 Differences of Accumulated Changes

This chapter helps you effectively use your calculator's numerical integrator with various applications pertaining to the accumulation of change. In this first section, we focus on the differences of accumulated changes.

7.1.1 FINDING THE AREA BETWEEN TWO CURVES
The process of finding the area of the region between two functions uses many of the techniques presented in preceding sections. If the two functions intersect, you need to first find the input values of the point(s) of intersection. We illustrate these ideas as they are presented in Example 2 of Section 7.1 of *Calculus Concepts*.

Clear all functions from the Y= list and turn off the stat plots. Enter the function $f(x) = 0.3x^3 - 3.3x^2 + 9.6x - 5.1$ in Y1 and enter the function $g(x) = -0.15x^2 + 2.03x - 5.1$ in Y2.	Plot1 Plot2 Plot3 \Y1■.3X^3-3.3X²+9.6X-5.1 \Y2■-.15X²+2.03X-5.1 \Y3= \Y4= \Y5=
From Figure 7.10 in the text, we see that the regions under discussion use Xmin = 1, Xmax = 7, Ymin ≈ −3.3, and Ymax ≈ 3.3. Press WINDOW, set these values, and then press GRAPH.	

We next find A and B, the inputs of the two points of intersection of the two functions. (These values will be the limits on the integrals we use to find the areas.) The method we use to find A is the intersection method that was discussed in Section 1.1.9 of this *Guide*.

With the graph on the screen, press 2nd TRACE (CALC) 5 [intersect]. The TI-83 asks FIRST CURVE? Press ENTER to mark Y1. The calculator then asks SECOND CURVE? Press ENTER to mark Y2. At the GUESS? prompt, use ◄ to move the cursor near the leftmost point of intersection.	Y2=-.15X²+2.03X-5.1 Guess? X=3.7446809 Y=.39830693
Press ENTER to find $A \approx 3.724$. To avoid making a mistake copying the x-value and to eliminate as much rounding error as possible, return to the home screen and store the x-value in A with the keystrokes X,T,θ,n STO▶ ALPHA MATH (A).	Intersection X=3.7238392 Y=.37934685
You could use this same method to find B, the input value of the rightmost point of intersection. However, to illustrate another way this can be done, we find B using the SOLVER. The equation we are solving is Y1 = Y2, but remember that what you enter in the SOLVER must have 0 on one side of the equation.	EQUATION SOLVER eqn:0=Y1-Y2

Copyright © Houghton Mifflin Company. All rights reserved.

Press ENTER. Note from the graph that B is near 7. Use this or some other value closer to B than it is to A as your guess in the SOLVER. Press ALPHA ENTER (SOLVE) to find that $B \approx 6.776$.

As with A, to avoid making a mistake copying this x-value and to eliminate as much rounding error as possible, return to the home screen and store this x-value in B with the keystrokes X,T,θ,n STO▶ ALPHA MATRX (B).

Calculate the areas of the three regions enclosed by f and g (that is, regions R1, R2, and R3 that are shown in Figure 7.10 in the text) as indicated to the right.

Note that each answer should be positive because we are finding areas. If you obtain a negative answer, you probably entered the functions in the wrong order. (Note that area R3 ≈ 0.166656.)

If you are in doubt as to which function to enter first in the integral, you can press GRAPH and trace the graphs to see which function is on top of the other. Press ▲ and ▼ to have the cursor jump from one curve to the other and notice the equation at the top of the screen.

Find $\int_1^7 [f(x) - g(x)]\,dx$ with the instruction shown to the right.

Note that the value of this integral is *not* the same as the total area between the two curves.

7.2 Streams in Business and Biology

You will find your calculator very helpful when dealing with streams that are accumulated over finite intervals. Finding either the future or present value of a continuous income stream is simply finding the value of a definite integral. However, the technique used with discrete income streams involves the sequence function that was introduced in Section 2.1.1. The details are presented in Section 7.2.2.

7.2.1 DETERMINING THE FLOW-RATE FUNCTION FOR AN INCOME STREAM

The TI-83 can often help you to find the equation for an income stream flow rate. Note that it is *not* necessary to use the calculator to find such an equation – we present this as a technique to use only if you find writing the equation from the word description difficult. We illustrate these ideas as they are given in Example 1 of Section 7.2 of *Calculus Concepts*.

In Example 1, part *a*, we are told that the business's profit remains constant. Clear lists L1 and L2. In L1 enter two possible input values for the time involved. (You might use different years than the ones shown here.) In L2 enter the amount invested: 10% of the constant profit.

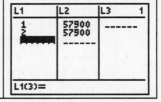

You need to remember that a *constant output* means a *linear flow rate*. Fit a linear function to these two data points to find that $R(t) = 57,000$ dollars per year.	LinReg y=ax+b a=0 b=57900

CAUTION: If you attempt to draw a scatter plot on the calculator, you will get an error message because the TI-83, using the output data in L2, sets Ymin = Ymax. (You need to draw the scatter plot using paper and pencil.) If you want to see the horizontal line graph on the TI-83, change Ymin and Ymax so that 57,900 is in between the two values and press GRAPH.

In Example 1, part *b*, we are told that the business's profit grows by $50,000 each year. The first year's profit (which determines the initial investment at $t = 0$) is $579,000. Reason that if the profit grows by $50,000 each year, the next year's profit will be $579,000 + 50,000 = $629,000. Enter these values in L1 and L2.	L1 / L2 / L3 0 / 579000 1 / 629000 L2(3) =
You need to remember that a *constant growth* means a *linear flow rate*. Fit a linear function to these two data points. Now, carefully read the problem once more. Note that only 10% of the profit is invested. Thus, the linear flow rate function is $R(t) = 0.10(50,000t + 579,000)$ dollars per year *t* years after the first year of business.	LinReg y=ax+b a=50000 b=579000
In Example 1, part *c*, we are told that the business's profit grows by 17% each year. The first year's profit (which determines the initial investment at $t = 0$) is $579,000. Reason that if the profit grows by 17% each year, the next year's profit will be $579,000 + 0.17(579,000) = $677,430. Enter these values in L1 and L2.	L1 / L2 / L3 0 / 579000 1 / 677430 L2(3) =
You need to remember that a *constant percentage growth* means an *exponential flow rate*. Fit an exponential function to these two data points. Now, carefully read the problem once more. Note that only 10% of the profit is invested. Thus, the exponential flow rate function is $R(t) = 0.10(579,000)(1.17^t)$ dollars per year *t* years after the first year of business.	ExpReg y=a*b^x a=579000 b=1.17

Part *d* of Example 1 gives data that describe the growth of the business's profit. Refer to the material in Section 2.1.4 of this *Guide* to review how to fit a log function to these data points.

Note: If you forget which type of growth gives which function, simply use what you are told in the problem and fill in the lists with approximately five data points. Draw a scatter plot and it should be obvious which function to fit to the data.

For instance, return to Example 1, part *b*, in which we are told that the business's profit grows by $50,000 each year. Note that we assume throughout this section that initial investments are made at time $t = 0$. Simply add $50,000 to the previous year's profit to obtain about five data points.	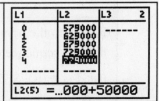

Draw a scatter plot, and observe that the points fall in a line! Note that if you run program DIFF, the first differences are constant at 50,000.

Next, find a linear equation for the data and find that the flow rate function is $R(t) = 0.10(50{,}000t + 579{,}000)$ dollars per year t years after the first year of business.

7.2.2 FUTURE VALUE OF A DISCRETE INCOME STREAM

The future value of a discrete income stream is found by adding, as d increases, the terms of $R(d)\left(1+\frac{r}{n}\right)^{D-d}$ where $R(d)$ is the value per period of the dth deposit, $100r\%$ is the annual percentage rate at which interest is earned when the interest is compounded once in each deposit period, n is the number of times interest is compounded (and deposits are made) during the year, and D is the total number of deposit periods. It is assumed that initial deposits are made at time $t = 0$ unless it is otherwise stated. We use the situation in Example 4 in Section 7.2 of *Calculus Concepts*:

> When you graduate from college (say, in 3 years), you would like to purchase a car. You have a job and can put $75 into savings each month for this purchase. You choose a money market account that offers an APR of 6.2% compounded quarterly.

Part *a* asks how much will you have deposited in 3 years. No interest is involved in this calculation. There are 12 months in each year and $75 is deposited each month for 3 years.

A total of $2700 is deposited.

```
12*3*75
            2700
```

Part *b* of Example 4 asks for the future value of the deposits at the end of 3 years. Because the APR is 6.2% compounded quarterly, $r = 0.062$, $n = 4$, and $D = 3$ years $\cdot\, 4\frac{\text{deposits}}{\text{year}} = 12$ deposits. A constant $75 each month is deposited, so $R(d) = \$75(3) = \$225$ deposited each quarter. The 3-year future value is given by $\sum_{d=0}^{11} 225\left(1+\frac{0.062}{4}\right)^{12-d}$.

The TI-83 sequence command finds this sum. The syntax for this sequence command is

seq(formula, variable, first value, last value, increment)

When applied to discrete income streams, the *formula* is $R(d)\left(1+\frac{r}{n}\right)^{D-d}$.

For convenience, we enter the formula in Y1. You must enclose the exponent in parentheses. Otherwise, the TI-83 will assume that only the 12 is in the exponent.

```
Plot1 Plot2 Plot3
\Y1■225(1+.062/4
)^(12-X)
\Y2=
\Y3=
\Y4=
\Y5=
\Y6=
```

Return to the home screen and enter seq(Y1, X, 0, 11, 1). Recall that the sequence command is accessed by pressing [2nd] [STAT] (LIST) ▶ [OPS] 5 [seq(]. If you want to see the 12 future values in this list, scroll through the list with ▶.

```
seq(Y1,X,0,11,1)
 {270.6086352 26…
```

NOTE: When you use the sequence command for discrete income streams, the first value is always 0. The last value is always $D - 1$ because we start counting at 0, not 1. The increment will always be 1 because of the way the formula is designed.

What are the values in the list that results when you use the sequence command? The first value (approximately $270.61) is the 3-year future value of the first 3 months deposits ($225). The second value (approximately $266.48) is the 3-year future value of the second 3 months of deposits ($225), and so forth.

To find the future value of all the deposits, simply sum the values in the list. To do this, use the sum command to add the values in the list (the previous ANS). The sum command is accessed with [2nd] [STAT](LIST) [▶] [▶] [MATH] 5 [sum(]. Store this value in F for use in the next part of the problem.	`seq(Y₁,X,0,11,1)` `{270.6086352 26…` `sum(Ans)` `       2988.101231` `Ans→F` `       2988.101231`
Part *c* of Example 4 asks for the present value of the amount you would have to deposit now to achieve the same future value as was found above. To find the present value P, solve the equation $P\left(1+\frac{r}{n}\right)^D =$ Future value. Enter $P\left(1+\frac{0.062}{4}\right)^{12} - F = 0$ in the SOLVER.	`EQUATION SOLVER` `eqn:0=P(1+.062/4` `)^12-F`
Press [ENTER]. Enter a guess for P, and do not change the value of F. (There is only one solution, so any positive number should work as a guess.) With the cursor on the P line, press [ALPHA] [ENTER] (SOLVE) to find $P \approx \$2484.48$..	`P(1+.062/4)^1…=0` `•P=2484.4838243…` `  F=2988.1012310…` `  bound={-1E99,1…` `•left-rt=0`
Part *d* of Example 4 asks for the future value if the monthly interest rate is 0.5% instead of the previous APR compounded quarterly. Press [Y=] and make the appropriate changes in the function in Y1. Don't forget that $R(d)$ is the amount deposited per interest compounding period.	`Plot1 Plot2 Plot3` `\Y₁■75(1+.005)^(` `36-X)` `\Y₂=` `\Y₃=` `\Y₄=` `\Y₅=` `\Y₆=`
Press [2nd] [ENTER] (ENTRY) until the sequence command reappears on the home screen. Change *last value* to 35 and press [ENTER]. Sum the resulting sequence to find that the 3-year future value is about $2964.96.	`seq(Y₁,X,0,35,1)` `{89.75103936 89…` `sum(Ans)` `       2964.958912`

NOTE: You can use any letter for the variable in the formula if you enter the function on the home screen, but if you are entering the formula in the Y= list, use X.

EXTENSION: The TI-83 calculator has a built-in financial menu accessed with [2nd] [x^{-1}] (FINANCE). The items in this menu can be extremely useful in other courses with discrete financial applications. Consult your *Owners' Manual* for more details and instructions.

7.3 Integrals in Economics

Consumers' surplus (when defined by a definite integral) and producers' surplus are easy to find using fnInt. You should draw graphs of the demand and supply functions and think of the economic quantities in terms of area to better understand the questions being asked.

7.3.1 CONSUMER ECONOMICS

We illustrate how to find the consumers' surplus and other economic quantities when the demand function intersects the input axis as given in Example 2 of Section 7.3 of *Calculus Concepts*:

> Suppose the demand for a certain model of minivan in the United States can be described as $D(p) = 14.12(0.933^p) - 0.25$ million minivans when the market price is p thousand dollars per minivan.

We first draw a graph of the demand function. This is not asked for, but it will really help your understanding of the problem. Read the problem to see if there are any clues as to how to set the horizontal view for the graph. The price cannot be negative, so $p \geq 0$. There is no price given in the remainder of the problem, so just guess a value with which to begin.

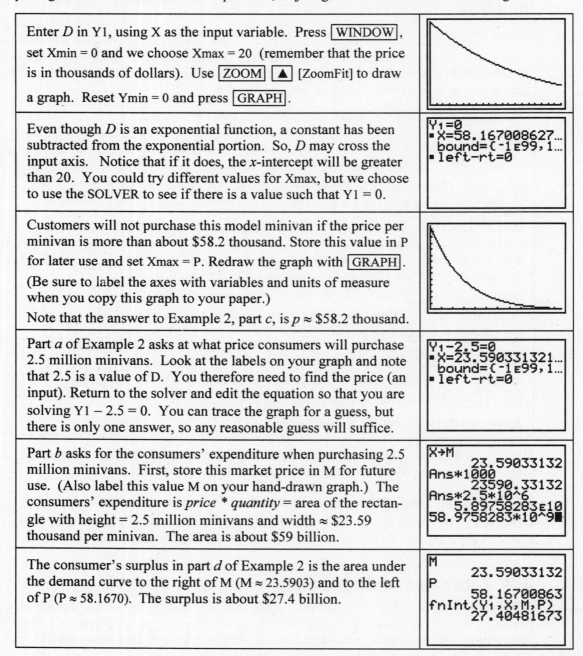

Enter D in Y1, using X as the input variable. Press [WINDOW], set Xmin = 0 and we choose Xmax = 20 (remember that the price is in thousands of dollars). Use [ZOOM] [▲] [ZoomFit] to draw a graph. Reset Ymin = 0 and press [GRAPH].	
Even though D is an exponential function, a constant has been subtracted from the exponential portion. So, D may cross the input axis. Notice that if it does, the x-intercept will be greater than 20. You could try different values for Xmax, but we choose to use the SOLVER to see if there is a value such that Y1 = 0.	
Customers will not purchase this model minivan if the price per minivan is more than about $58.2 thousand. Store this value in P for later use and set Xmax = P. Redraw the graph with [GRAPH]. (Be sure to label the axes with variables and units of measure when you copy this graph to your paper.) Note that the answer to Example 2, part *c*, is $p \approx \$58.2$ thousand.	
Part *a* of Example 2 asks at what price consumers will purchase 2.5 million minivans. Look at the labels on your graph and note that 2.5 is a value of D. You therefore need to find the price (an input). Return to the solver and edit the equation so that you are solving Y1 − 2.5 = 0. You can trace the graph for a guess, but there is only one answer, so any reasonable guess will suffice.	
Part *b* asks for the consumers' expenditure when purchasing 2.5 million minivans. First, store this market price in M for future use. (Also label this value M on your hand-drawn graph.) The consumers' expenditure is *price * quantity* = area of the rectangle with height = 2.5 million minivans and width ≈ \$23.59 thousand per minivan. The area is about \$59 billion.	
The consumer's surplus in part *d* of Example 2 is the area under the demand curve to the right of M (M ≈ 23.5903) and to the left of P (P ≈ 58.1670). The surplus is about \$27.4 billion.	

TI-83 Guide   A-81

The TI-83 draws the consumers' surplus if you use the methods illustrated in Section 6.5.2.

First, reset Ymin to −2 to have more room at the bottom of the graph. Draw the graph of *D*. Then, with the graph on the screen, press [2nd] [TRACE] (CALC) 7 [∫f(x)dx]. At the LOWER LIMIT? prompt, press [ALPHA] [÷] (M) [ENTER]. At the UPPER LIMIT? prompt, press [ALPHA] 8 (P) [ENTER]. You should then see the screen shown to the right.	

CAUTION: Unless the values you enter for the upper and lower limits are visible on the graphics screen, you will get an error message using the above instructions. If you do as instructed and use P, make sure you used P or a value larger than P when setting Xmax.

7.3.2 PRODUCER ECONOMICS

We illustrate how to find the producers' surplus and other economic quantities as given in Example 4 of Section 7.3 of *Calculus Concepts*:

Suppose the demand and supply functions for a certain model minivan are described by $D(p) = 14.12(0.933^p) - 0.25$ million minivans and $S(p) = 0$ minivans for $p < 15$ and $S(p) = 0.25p - 3.75$ million minivans for $p \geq 15$ when the market price is p thousand dollars per minivan.

We first draw graphs of the demand and supply functions.

Enter *D* in Y1. Enter S (for $p \geq 15$) in Y2 as shown to the right. Use X as the input variable in each function. Recall that you need to enclose the piecewise function and its input in parentheses. Access the inequality symbol with [2nd] [MATH] 4 [≥].	
Press [WINDOW], set Xmin = 0 and we choose Xmax = 58.167. (This is the approximate value at which *D* = 0, as determined in the previous section.) Use [ZOOM] [▲] [ZoomFit] to draw the graphs of *D* and *S*. If Ymin is not 0, set it to 0 and press [GRAPH].	
Part *a* of Example 4 asks for the market equilibrium point. You can find this by using the SOLVER (solve Y1 − Y2 = 0) or by using the intersection method on the graph. We show the latter. See Section 7.1.1 for the specifics of using this method. The *x*- and *y*-values displayed at the bottom of the screen are the location of the market equilibrium point. Store this X-value in E.	
The total social gain is producers' surplus + consumers' surplus at the equilibrium price. Shade this area on your graph and then use the graph to write the integrals that give this area: Social gain = $\int_{15}^{E} S(p)\,dp + \int_{E}^{P} D(p)\,dp$ where E ≈ 24.3996 and P ≈ 58.1670. (The value of P was found in Section 7.3.1.)	

7.4 Average Value and Average Rates of Change

Average rates of change are computed as discussed in Section 3.1.1 of this *Guide*. When finding an average value, you need to carefully read the question in order to determine which quantity should be integrated. Considering the units of measure in the context can be a great help when trying to determine which function to integrate to find an average value.

7.4.1 AVERAGE VALUE OF A FUNCTION We illustrate finding an average value with the data in Table 7.18 in Example 1 of Section 7.4 of *Calculus Concepts*:

Time (number of hours after midnight)	7	8	9	10	11	12	13	14	15	16	17	18	19
Temperature (°F)	49	54	58	66	72	76	79	80	80	78	74	69	62

Clear any old data. Delete any functions in the Y= list and turn on Plot 1. Enter the data in the above table in lists L1 and L2.

A scatter plot of the data indicates an inflection point (around 9 p.m.) and no limiting values.

Fit a cubic function to the data and paste it in Y1.

Part *b* of Example 1 asks for the average *temperature* (*i.e.*, the average value of the temperature) between 9 a.m. and 6 p.m. So, integrate the *temperature* between $x = 9$ and $x = 18$ and divide by the length of the interval to find that the answer is about 74.4°F.

7.4.2 GEOMETRIC INTERPRETATION OF AVERAGE VALUE What does the average value of a function mean in terms of the graph of the function? We continue with Example 1 of Section 7.4 of *Calculus Concepts* by considering the function and average value found in Section 7.4.1. Have the unrounded cubic temperature function in Y1.

Enter 74.42722902 in Y2. You can either leave Ymin at the setting from the scatter plot or change Ymin to 0. Press GRAPH to see the graph of the average value and the function.

We illustrate for only the second graph, but what follows is true for both. The area of the rectangle whose height is the average temperature is $(74.42722902 - 0)(18 - 9) \approx 669.845$. The area of the region between the temperature function and the input axis between 9 a.m. and 6 p.m. is this same value.

To find the answer to part *d* of Example 1, enter and evaluate the expression shown to the right. The average rate of change of the temperature from 9 a.m. to 6 p.m. is about 0.98°F/hr.

NOTE: The material in the rest of this *Guide* appears only in the complete version of the text.

7.5 Probability Distributions and Density Functions

Most of the applications of probability distributions and density functions use technology techniques that have already been discussed. Probabilities are areas whose values can be found by integrating the appropriate density function. A cumulative density function is an accumulation function of a probability density function.

Your calculator's numerical integrator is especially useful for finding means and standard deviations of some probability distributions because those integrals often involve expressions for which we have not developed algebraic techniques for finding antiderivatives. The TI-83 calculator contains many built-in statistical menus that deal with probability distributions and density functions. These are especially useful when you take a statistics course, but some can be used with Section 7.5 in this course. Consult your TI-83 *Owners' Manual* for more details and instructions for use.

7.5.1 NORMAL PROBABILITIES The normal density function is the most well known and widely used probability distribution. If you are told that a random variable x has a normal distribution $N(x)$ with mean μ and standard deviation σ, the probability that x is between two real numbers a and b is given by

$$\int_a^b N(x)\,dx = \int_a^b \frac{1}{\sigma\sqrt{2\pi}} e^{\frac{-(x-\mu)^2}{2\sigma^2}}\,dx$$

We illustrate these ideas with the situation in Example 8 of Section 7.5 of *Calculus Concepts*. In that example, we are given that the distribution of the life of the bulbs, with the life span measured in hundreds of hours, is modeled by a normal density function with $\mu = 900$ hours and $\sigma = 100$ hours. Carefully use parentheses when entering the normal density function.

The probability that a light bulb lasts between 9 and 10 hundred hours is $$P(9 \leq x \leq 10) = \int_9^{10} \frac{1}{\sqrt{2\pi}} e^{-(x-9)^2/2}\,dx$$ The value of this integral must be a number between 0 and 1.	`Plot1 Plot2 Plot3`
	`\Y1■1/√(2π)e^(-(`
	`X-9)²/2)`
	`\Y2=`
	`\Y3=`
	`\Y4=`
	`\Y5=`
	`\Y6=`
		`fnInt(Y1,X,9,10)`
		`      .3413447461`

7.5.2 VIEWING NORMAL PROBABILITIES We could have found the probability that the light bulb life is between 900 and 1000 hours graphically by using the CALC menu.

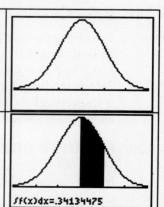

To draw a graph of the normal density function, note that nearly all of the area between this function and the horizontal axis lies within three standard deviations of the mean. Set Xmin = 9 − 3 = 6 and Xmax = 9 + 3 = 12. Use ZoomFit to draw the graph. Reset Ymin to −0.1 in preparation for the next step, and press GRAPH.

With the graph on the screen, press 2nd TRACE (CALC) 7 [∫f(x)dx]. At the Lower Limit? prompt, enter 9. At the Upper Limit? prompt, enter 10. The probability is displayed as an area and the value is printed at the bottom of the graphics screen.

Chapter 8 Repetitive Change: Cycles and Trigonometry

8.1 Functions of Angles: Sine and Cosine

Before you begin this chapter, go back to the first page of this *Guide* and check the basic setup, the statistical setup, and the window setup. If these are not set as specified in Figures 1, 2, and 3, you will have trouble using your calculator in this chapter. Pay careful attention to the third line in the MODE screen in the basic setup. The Radian/Degree mode setting affects the TI-83's interpretation of the ANGLE menu choices. The calculator's MODE menu should always be set to Radian unless otherwise specified. (Note that calculator instructions for material that is in the *Trigonometry Appendix* at the *Calculus Concepts* Website is on page A-105 of this *Guide*.)

8.1.1 FINDING OUTPUTS OF TRIG FUNCTIONS WITH RADIAN INPUTS It is essential that you have the correct mode set when evaluating trigonometric function outputs. The angle setting in the MODE menu must be Radian for all applications in Chapter 8. We show how to evaluate trig functions as presented in Example 2 of Section 8.1 of *Calculus Concepts*.

Find $\sin \frac{9\pi}{8}$ and $\cos \frac{9\pi}{8}$. Because these angles are in radians, be certain that Radian is chosen in the third line of the MODE screen. The [SIN] and [COS] keys are above the [,] and [(] keys on the TI-83 keyboard.	sin(9π/8)
	   -.3826834324
	cos(9π/8)
	   -.9238795325

It is also essential that you use parentheses to indicate the order of operations. When you press any of the trig function keys, the left parenthesis automatically appears and cannot be deleted. If anything follows the angle, the right parenthesis is necessary to show the end of the input of the trig function.	sin(9π/8
	   -.3826834324
	sin(9π/8)+3
	   2.617316568
	sin(9π/8+3
	   .2484758395

Because the cosine and the sine values are, respectively, the x- and y- coordinates of the point where the terminal side of the angle $\frac{9\pi}{8}$ intersects the unit circle, these values satisfy the equation $x^2 + y^2 = 1$. The fact that $(\cos \theta)^2 + (\sin \theta)^2 = 1$ is true for any angle θ.	(cos(9π/8))²+(si n(9π/8))² 1

8.2 Cyclic Functions as Models

We now introduce another model – the sine model. As you might expect, this function should be fit to data that repeatedly varies between alternate extremes. The form of the sine model is given by $f(x) = a \sin(bx + h) + k$ where $|a|$ is the amplitude, b is the frequency (where $b > 0$), $2\pi/b$ is the period, $|h|/b$ is the horizontal shift (to the right if $h < 0$ and to the left if $h > 0$), and k is the vertical shift (up if $k > 0$ and down if $k < 0$). *Note:* The TI-83 uses the c when we use h and d when we use k.

8.2.1 FITTING A SINE MODEL TO DATA Before fitting any model to data, remember that you should construct a scatter plot of the data and observe what pattern the data appear to follow. Example 2 in Chapter 8 asks you to find a sine model for cyclic data with the hours of daylight on the Arctic Circle as a function of the day of the year on which the hours of daylight are measured. (January 1 is day 1.) These data appear in Table 8.3 of *Calculus Concepts*.

Day of the year	−10	81.5	173	264	355	446.5	538	629	720	811.5
Hours of daylight	0	12	24	12	0	12	24	12	0	12

Clear any old data. Delete any functions in the Y= list and turn on Plot 1. Enter the data in the above table in lists L1 and L2. Construct a scatter plot of the data. When using the sine regression in the TI-83, it is sometimes necessary to have an estimate of the period of the data.

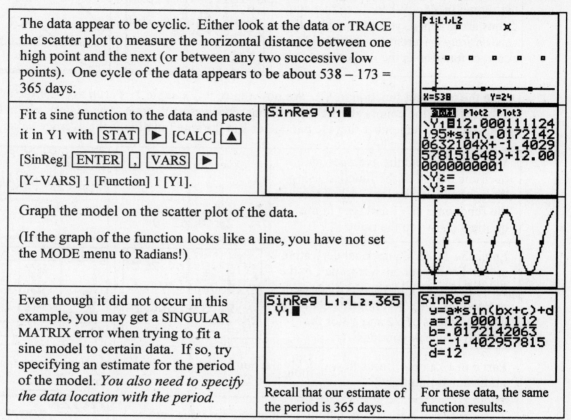

The data appear to be cyclic. Either look at the data or TRACE the scatter plot to measure the horizontal distance between one high point and the next (or between any two successive low points). One cycle of the data appears to be about 538 − 173 = 365 days.

Fit a sine function to the data and paste it in Y1 with [STAT] [▶] [CALC] [▲] [SinReg] [ENTER] [,] [VARS] [▶] [Y-VARS] 1 [Function] 1 [Y1].

Graph the model on the scatter plot of the data.

(If the graph of the function looks like a line, you have not set the MODE menu to Radians!)

Even though it did not occur in this example, you may get a SINGULAR MATRIX error when trying to fit a sine model to certain data. If so, try specifying an estimate for the period of the model. *You also need to specify the data location with the period.*

Recall that our estimate of the period is 365 days.

For these data, the same function results.

NOTE: If you do not think the original function the calculator finds fits the data very well, try specifying a period and see if a better-fitting equation results. It didn't here, but it might with a different set of data. If you still cannot find an equation, you can tell the TI-83 how many times to go through the routine that finds the equation. This number of iterations is 3 if not specified. The number should be typed before L1 when initially finding the equation.

8.3 Rates of Change and Derivatives

All the previous techniques given for other functions also hold for the sine model. You can find intersections, maxima, minima, inflection points, derivatives, integrals, and so forth.

8.3.1 DERIVATIVES OF SINE AND COSINE FUNCTIONS
Evaluate nDeriv at a particular input to find the value of the derivative of the sine function at that input. We illustrate with Example 1 in Section 8.3 of *Calculus Concepts*:

> The calls for service made to a county sheriff's department in a certain rural/suburban county can be modeled as $c(t) = 2.8 \sin(0.262t + 2.5) + 5.38$ calls during the tth hour after midnight.

CAUTION: Because the TI-83 automatically inserts a left parentheses when you press the sine and cosine keys, be very careful that you do not type an extra parentheses when entering a trig function in the graphing list. If you do, you will change the order of operations and not have the correct equation entered. The function (see below) that is entered in Y2 is <u>incorrect</u>.

The correct way to enter the function c is shown in Y1. When these two equations are graphed, you can see that they are entirely different functions! The correct function is the darker one.		

Delete the incorrect function in Y2. We now return to Example 1. Part *a* asks for the average number of calls the county sheriff's department receives each hour. The easiest way to obtain this answer is to remember that the parameter k in the sine function is the average value.

You can also find the average value over one period of the function using the methods of Section 7.4.1. Enter the function c in Y1 and type in the quotient shown to the right.	Average Value = $$\frac{\int_0^{2\pi/0.262} c(t)\,dt}{2\pi/0.262 - 0}$$	fnInt(Y₁,X,0,2π/.262)
		         129.0211334
		Ans/(2π/.262)
		                  5.38

Enter the TI-83's numerical derivative in Y2. Enter your answer to part *b*, the formula for $c'(t)$, in Y3. As discussed in Sections 4.4.3/4.4.4 of this *Guide*, use the graphs of Y2 and Y3 or the TABLE to check your answer.	Plot1 Plot2 Plot3 \Y₁=2.8sin(.262X+2.5)+5.38 \Y₂=nDeriv(Y₁,X,X) \Y₃=.7336cos(.262X+2.5) \Y₄=	X Y₂ Y₃ 0 -.5877 -.5877 2 -.7285 -.7285 5 -.5757 -.5757 10 .43691 .43691 22 -.2924 -.2924 34 .29405 .29405 X=

Part *c* of Example 1 asks how quickly the number of calls received each hour is changing at noon and at midnight.

To answer these questions, simply evaluate Y2 (or your derivative in Y3) at 12 for noon and 0 (or 24) for midnight. (You were not told if "midnight" refers to the initial time or 24 hours after that initial time.)	Y₂(12)
	    .588774165
	Y₂(0)
	   -.5877189497
	Y₂(24)
	   -.5898259682

8.5 Accumulation in Cycles

As with the other functions we have studied, applications of accumulated change with the sine and cosine functions involve the calculator's numerical integrator fnInt.

8.5.1 INTEGRALS OF SINE AND COSINE FUNCTIONS

We illustrate with Example 1 in Section 8.5 of *Calculus Concepts*.

Enter the rate of change of temperature in Philadelphia on August 27, 1993 in Y1. Find the accumulated change in the temperature between 9 a.m. and 3 p.m. using fnInt.	Plot1 Plot2 Plot3 \Y₁=2.733cos(.285X-2.93) \Y₂= \Y₃= \Y₄= \Y₅= \Y₆=	fnInt(Y₁,X,9,15) 12.76901162

Chapter 9 Ingredients of Multivariable Change: Models, Graphs, Rates

9.1 Cross-Sectional Models and Multivariable Functions

For a multivariable function with two input variables, obtain a cross-sectional model by entering the data in lists L1 and L2 and then fitting the appropriate function as indicated in previous chapters of this *Guide*. Unless you are told otherwise, we assume that the data are given in a table with the values of the *first* input variable listed *horizontally* across the top of the table and the values of the *second* input variable listed *vertically* down the left side of the table.

9.1.1 FINDING A CROSS-SECTIONAL MODEL FROM DATA (HOLDING THE FIRST INPUT VARIABLE CONSTANT) Using the elevation data in Table 9.2 of Section 9.1 in *Calculus Concepts*, find the cross-sectional model $E(0.8, n)$ as described below. Remember that "rows" go from left to right horizontally and "columns" go from top to bottom vertically.

n (miles)	1.5	1.4	1.3	1.2	1.1	1.0	0.9	0.8
Elevation (feet above sea level)	797.6	798.1	798.5	798.9	799.2	799.5	799.7	799.9
n (miles)	0.7	0.6	0.5	0.4	0.3	0.2	0.1	0.0
Elevation (feet above sea level)	800.0	800.1	800.1	800.1	800.0	799.9	799.7	799.5

So, refer for a moment to Table 9.1. When you are asked to find $E(0.8, n)$, e is constant at 0.8 and n varies. Thus, enter the values for n appearing on the left side of the table (vertically) in L1 and the elevations E obtained in the $e = 0.8$ column of Table 9.1 in L2. *The outputs that you enter in L2 will always be in the main body of any multivariable data table given in this text.*

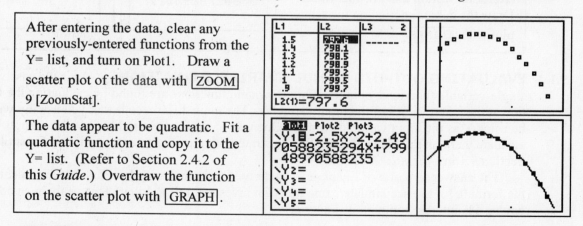

| After entering the data, clear any previously-entered functions from the Y= list, and turn on Plot1. Draw a scatter plot of the data with ZOOM 9 [ZoomStat]. |
| The data appear to be quadratic. Fit a quadratic function and copy it to the Y= list. (Refer to Section 2.4.2 of this *Guide*.) Overdraw the function on the scatter plot with GRAPH. |

CAUTION: Because you will often be asked to find several different cross-sectional models using the same data table, calling different variables by the same names x and y would be very confusing. It is very important that you call the variables by the names that have been assigned in the problem. Remember that when finding or graphing a function, the TI-83 always calls the input variable X and the output variable Y. When working with multivariable functions, you must translate the calculator's equation Y1 ≈ −2.5X^2 + 2.497X + 799.490 into the symbols that are used in the application. You should write the cross-sectional function as $E(0.8, n) = -2.5n^2 + 2.497n + 799.490$. Don't forget to completely describe (including units) all of the variables.

9.1.2 FINDING A CROSS-SECTIONAL MODEL FROM DATA (HOLDING THE SECOND INPUT VARIABLE CONSTANT)

The only difference in this and the previous section of this *Guide* is that the second input, instead of the first, is held constant. We illustrate using the situation in Example 1 of Section 9.1 in *Calculus Concepts*. Refer to Table 9.1. Because we are asked to find the cross-sectional model $E(e, 0.6)$, $n = 0.6$ and the inputs are the values of e that are across the top of the table. Enter these values in L1. (See below for a shortcut.) The outputs are the elevations E obtained in the $n = 0.6$ mile row in Table 9.1. Enter these in L2.

NOTE: You may find it helpful to place a piece of paper or a ruler under the row (or to the right of the column) in which the data appear to help avoid typing an incorrect value.

Because the input values for this function are the same as the input values in the last section of this *Guide* (but in reverse order) you can sort L1 in ascending order to avoid re-entering the data. Press [2nd] [STAT] (LIST) ▶ [OPS] 1 [SortA(] [2nd] 1 (L1) [ENTER]. Next, enter the output in L2.	SortA(L1) Done	
After entering the data, clear any functions from the Y= list, and turn on Plot1. Draw a scatter plot of the data with [ZOOM] 9 [ZoomStat].	L1: 0,.1,.2,.3,.4,.5,.6 L2: 802.8, 801.6, 800.8, 800.3, 800, 799.9, 799.9 L2(1)=802.8	(scatter plot)
There is an inflection point and no evidence of limiting values, so the data appear to be cubic. Fit a cubic function and copy it to the Y= list. (Refer to Section 2.4.3 of this *Guide*.) Draw the cubic function on the scatter plot with [GRAPH].	CubicReg $y=ax^3+bx^2+cx+d$ $a=-10.12410362$ $b=21.34716032$ $c=-13.97232949$ $d=802.8093911$	(cubic fit on scatter plot)

9.1.3 EVALUATING OUTPUTS OF MULTIVARIABLE FUNCTIONS

As is the case with single-variable functions, outputs of multivariable functions are found by evaluating the function at the given values of the input variables. The main difference is that you usually will not be using X as the input variable symbol. One way to find multivariable function outputs is to evaluate them on the home screen. We illustrate with the investment function in Example 2 of Section 9.1 in *Calculus Concepts*.

The answer to part *a* of Example 2, as derived from the compound interest formula, uses the formula for the accumulated amount of an investment of P dollars for t years in an account paying 6% interest compounded quarterly:

$$A(P, t) = P\left(1 + \frac{0.06}{4}\right)^{4t} \text{ dollars}$$

When 10 is substituted for t, the cross-sectional function becomes $A(P, 10) \approx 1.814018409P$.

Part *b* of Example 2 asks for a model for the data in the row of Table 9.4 that corresponds to 10 years. Thus, we want to find $A(P, 10)$. Enter the data from the table in L1 and L2.

TI-83 Guide

| A scatter plot of the data appears to be linear. Program DIFF confirms that a linear function is appropriate because the first differences are constant. Fit a linear function and paste it in Y1. | L1: 3500, 4000, 4500, 5000, 5500, 6000, 6500 / L2: 6349.1, 7256.1, 8163.1, 9070.1, 9977.1, 10884, 11791 / L3: 907.01, 907.01, 907.01, 907.01, 907.01, 907.01
 L3(1)=907.01 | Plot1 Plot2 Plot3
 \Y1■1.81402X+-.01
 \Y2=
 \Y3=
 \Y4=
 \Y5=
 \Y6= |

Notice that the cross-sectional functions in parts *a* and *b* are slightly different because the data in Table 9.4 have been rounded to the nearest cent. Part *c* of Example 2 asks for $A(5300, 10)$. Even though it is simplest here to substitute 5300 for P in $A(P, 10) \approx 1.814018409P$, we return to the original function to illustrate evaluating multivariable formulas on the calculator.

| To find the output on the home screen, type the formula for $A(P, t)$, substituting $P = 5300$ and $t = 10$. Press ENTER.

 Again, be warned that you must carefully use the correct placement of parentheses. Check your result with the values in Table 9.4 in the text to see if \$9614.30 is a reasonable amount. | 5300(1+.06/4)^(4*10)
                9614.297566 |

Even though it is not necessary in this example, you may encounter activities in this section in which you need to evaluate a multivariable function at several different inputs. You could use what is shown above, but there are easier methods than entering each one individually. You will also use the techniques shown below in later sections of this chapter. When evaluating a multivariable function at several different input values, you may find it more convenient to enter the multivariable function in the graphing list.

Clear any previously-entered equations. Enter in Y1 the function $A(P, t) = P\left(1 + \frac{0.06}{4}\right)^{4t}$ with the keystrokes ALPHA 8 (P) (1 + . 06 ÷ 4) ^ (4 ALPHA 4 (T)).	Plot1 Plot2 Plot3 \Y1■P(1+.06/4)^(4T) \Y2= \Y3= \Y4= \Y5= \Y6=

Return to the home screen and input the values $P = 5300$ and $t = 10$ with the keystrokes 5300 STO▶ ALPHA 8 (P) ALPHA . (:) 10 STO▶ ALPHA 4 (T) ENTER. Evaluate Y1 at these inputs with VARS ▶ [Y-VARS] 1 [Function] 1 [Y1] ENTER.	5300→P:10→T
	                10
	 Y1
	         9614.297566

To evaluate Y1 at other inputs, press 2nd ENTER (ENTRY) twice to recall the storing instruction. Change the values and press ENTER. Then press 2nd ENTER (ENTRY) twice to recall Y1 and press ENTER.	10
	 Y1
	         9614.297566
	 6500→P:8→T
	                8
	 Y1
	         10467.10808

CAUTION: It is very important to note at this point that while we have previously used X as the input variable when entering functions in the Y= list, we do *not* follow this rule when we *evaluate* functions with more than one input variable. However, realize that we should not graph Y1 nor use the TABLE. If you attempt to graph the current Y1 = $A(P, t)$ or use the table, you will see that the calculator considers this Y1 a constant. (Check it out and see that the graph is a horizontal line at about 10467.1 and that all values in the table are about 10467.1.)

Copyright © Houghton Mifflin Company. All rights reserved.

9.2 Contour Graphs

Because any program that we might use to graph[1] a three-dimensional function would be fairly involved and take a long time to execute, we do not graph three-dimensional functions. Instead, we discover information about three-dimensional graphs using their associated contour curves.

9.2.1 SKETCHING CONTOUR CURVES When given a multivariable function with two input variables, you can draw contour graphs using the three-step process described below. We illustrate with the function that gives the monthly payments required to pay off a loan of A thousand dollars over a period of t years at 7% interest. This function, M, appears in Example 1 of Section 9.2 of *Calculus Concepts*:

$$M(t, A) = \frac{5.833333A}{1 - 0.932583^t} \text{ dollars}$$

Step 1: We are told that the monthly payment on the loan is $520. Set $M(t, A) = 520$. Because we will use the solve routine to find the value of A at various values of t, write the function as $M(t, A) - 520 = 0$.

Step 2: Choose values for t and solve for A to obtain points on the $520 constant-contour curve. Obtain guesses for the values of t and A from the table of loan amounts in Figure 9.15 of the text.

Enter the $M(t, A)$ formula in Y1. Be sure that you use the letters A and T and not X. (We are only using Y1 as a "holding place" for the formula.) Go to the SOLVER and enter Y1 – 520 after eqn:0 =. Press ENTER.	Plot1 Plot2 Plot3 \Y1■5.833333A/(1 -.932583^T) \Y2= \Y3= \Y4= \Y5= \Y6=	EQUATION SOLVER eqn:0=Y1-520
Use ▼ to move to the T line and type 12. Press ▲ to move the cursor to the A line – the cursor must be on the line of the unknown variable when you press ALPHA ENTER (SOLVE).	Y1-520=0 A=50■ T=12 bound={-1E99,1... left-rt=0	Y1-520=0 ■A=50.565146552... T=12 bound={-1E99,1... ■left-rt=0
Enter another value for t, say $t = 16$ years. Move the cursor to the A line and press ALPHA ENTER (SOLVE) to find $A \approx \$59.963$ thousand.	Y1-520=0 ■A=59.962781721... T=16 bound={-1E99,1... ■left-rt=0	Y1-520=0 ■A=67.071127719... T=20 bound={-1E99,1... ■left-rt=0
Repeat the procedure for $t = 20$, $t = 24$, $t = 28$, and $t = 30$. Make a table of the values of t and A as you find them.	Y1-520=0 ■A=76.514808554... T=28 bound={-1E99,1... ■left-rt=0	Y1-520=0 ■A=78.160104385... T=30 bound={-1E99,1... ■left-rt=0

---

[1] There are many TI-83 programs, including one that will graph a multivariable function with two input variables, available at the web site with address http://education.ti.com/product/tech/83/down/download.html.

Step 3: Plot the points obtained in Step 2 with pencil and paper. You need to find as many points as it takes to see the pattern the points are indicating when you plot them. Connect the points with a smooth curve.

You need a function to draw a contour graph using the above method. Even though there may be several functions that seem to fit the data points obtained in Step 2, their use would be misleading because the real best-fit function can only be determined by substituting the appropriate values in a multivariable function. The focus of this section is to use contour graphs to study the relationships between input variables, not to find the equation of a function to fit a contour curve. Thus, we always sketch the contours on paper rather than with the TI-83.

9.3 Partial Rates of Change

When you hold all but one of the input variables in a multivariable function constant, you are actually looking at a function of one input variable. Thus, every technique we discussed previously can be used. In particular, the calculator's numerical derivative nDeriv can be used to find partial rates of change at specific values of the varying input variable. Although your TI-83 does not give formulas for derivatives, you can use it as discussed in Sections 4.4.4 and/or 4.4.4 of this *Guide* to check your algebraic formula for a partial derivative.

9.3.1 NUMERICALLY CHECKING PARTIAL DERIVATIVE FORMULAS As mentioned in Chapter 4, the basic concept in checking your algebraically-found partial derivative formula is that your formula and the calculator's formula computed with nDeriv should have the same outputs when each is evaluated at several different randomly-chosen inputs. You can use the methods in Section 9.1.3 of this *Guide* to evaluate each derivative formula at several different inputs and determine if the same numerical values are obtained from each formula. We illustrate these ideas by checking the answers for the partial derivative formulas found in parts *b* and *d* of Example 1 for this investment function:

The accumulated value of an investment of *P* dollars over *t* years at an APR of 6% compounded quarterly is $A(P, t) = P(1.061363551^t)$.

Recall that the syntax for the calculator's numerical derivative is

nDeriv(function, symbol for input variable, point at which the derivative is evaluated)

Enter the function *A*, using the letters P and T that appear in the formula, in Y1.
 	
Part *b* of Example 1 asks for a formula for $\partial A/\partial t$, so enter your formula (which may not be the same as the one that is shown to the right) in Y2. Enter the TI-83's derivative, using T as the changing input, in Y3.
	`Plot1 Plot2 Plot3`
	`\Y1=P(1.06136355`
	`1)^T`
	`\Y2=Pln(1.061363`
	`551)(1.061363551`
	`)^T`
	`\Y3=nDeriv(Y1,T,`
	`T)`

Store a value in T and call up Y2 and Y3. Do this for several different inputs. If the outputs of Y2 and Y3 are the same, your answer in Y2 is *probably* correct.	`5→T:Y2`
	`     521.3728632`
	`Y3`
	`     521.3728637`
	`10→T:Y2`
	`     702.2136524` `Y3`
	`     702.2136524`
	`     702.2136525`
	`25→T:Y2`
	`     1715.662291`
	`Y3`
	`     1715.662292`

NOTE: Remember that [2nd] [ENTER] (ENTRY) recalls previously-entered statements so you can save time. Also recall that [ALPHA] [.] (:) joins two or more statements together. Only the value of the last statement is printed, so do not join all three together with this symbol.

You may find it more convenient to numerically check your answer using the TABLE. If so, you must remember that when using the TABLE, the TI-83 considers X as the variable that is changing. When finding a partial derivative formula, all other variables are held constant except the one that is changing. So, to use the TABLE (or draw a graph), just store values in all constants and call the changing variable X. Then, proceed according to the directions given in Chapter 4. We illustrate this idea with part *d* of Example 1.

Have the function A, using the letters P and T that appear in the formula, in Y1. Because P is the variable that is changing in $\partial A/\partial P$, replace P with X. Turn off Y1.
 	
Enter your formula for $\partial A/\partial P$ (which may not be the same as the one that is shown to the right) in Y2. Enter the TI-83's derivative, using X as the changing input, in Y3.
	`Plot1 Plot2 Plot3`
	`\Y1=X(1.06136355`
	`1)^T`
	`\Y2=(1.061363551`
	`)^T`
	`\Y3=nDeriv(Y1,X,X)`
	`\Y4=`

Return to the home screen. Because T is constant, enter any reasonable value, except 0, in T. (You could repeat what comes next for T = 2, 10, and 37 as shown in part *c* of Example 1.) (If you need additional instructions on using the TABLE, see Section 1.1.6 of this *Guide*.)	`2→T` `              2`

Have TBLSET set to ASK in the Indpnt: location. Go to the TABLE and input some reasonable values for t = X.
 	
If the values shown for Y2 and Y3 are the same, your formula in Y2 for $\partial A/\partial P$ is *probably* correct.
	`TABLE SETUP`
	`TblStart=0`
	`ΔTbl=1`
	`Indpnt: Auto Ask`
	`Depend: Auto Ask`
	
	`X  Y2      Y3`
	`0  1.1265  1.1265`
	`2  1.1265  1.1265`
	`5  1.1265  1.1265`
	`10 1.1265  1.1265`
	`22 1.1265  1.1265`
	`34 1.1265  1.1265`
	`Y2=1.12649258739`

Did you wonder why the values in Y2 and Y3 are the same for all values of T? (It is because the function in Y1 is a linear function with a constant slope. Remember that T is a constant because P is the changing variable in $\partial A/\partial P$.)

NOTE: If you have an unrounded function that was fit to data in Y1 and you use the rounded function to compute your partial derivative, the Y2 and Y3 columns will be slightly different.

9.3.2 VISUALIZING AND ESTIMATING PARTIAL RATES OF CHANGE A partial rate of change of a multivariable function (evaluated at a specific point) is the slope of the line tangent to the graph of a cross-sectional model at a given location. We illustrate this concept in this section and in Section 9.3.3 using the Missouri farmland cross-sectional equations for elevation: $E(0.8, n)$ and $E(e, 0.6)$. It would be best to use the unrounded functions that were found in Sections 9.1.1 and 9.1.2 of this *Guide*. However, here we use the rounded functions rather than re-enter all the data.

Enter $-2.5n^2 + 2.497n + 799.490 = E(0.8, n)$ in Y1. Because we are going to graph this function, use X, not n, as the input variable.
 	
Press WINDOW and set values such as those shown to the right.
	`Plot1 Plot2 Plot3`
	`\Y1=-2.5X²+2.497`
	`X+799.490`
	`\Y2=`
	`\Y3=`
	`\Y4=`
	`\Y5=`
	`\Y6=`
		`WINDOW`
		`Xmin=0`
		`Xmax=1.5`
		`Xscl=1`
		`Ymin=795`
		`Ymax=804`
		`Yscl=1`
		`Xres=1`

The window settings can be obtained by drawing a scatter plot of the data used to obtain the function $E(0.8, n)$ or by looking at the $e = 0.8$ column in Table 9.1 in *Calculus Concepts*.

You can draw the line tangent to the graph at $n = X = 0.6$ from either the home screen or the graphics screen. (See Section 3.2.2 of this *Guide* for an explanation of both methods.) We use the home screen in this section and the graphics screen in the next section.

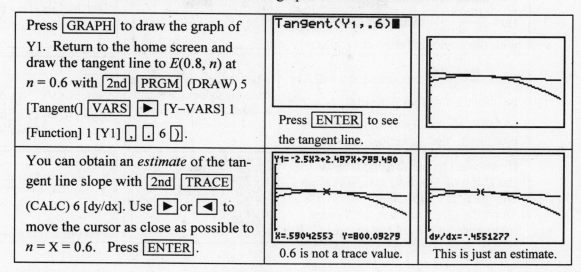

Press GRAPH to draw the graph of Y1. Return to the home screen and draw the tangent line to $E(0.8, n)$ at $n = 0.6$ with 2nd PRGM (DRAW) 5 [Tangent(] VARS ▶ [Y-VARS] 1 [Function] 1 [Y1] , . 6) .	Press ENTER to see the tangent line.	
You can obtain an *estimate* of the tangent line slope with 2nd TRACE (CALC) 6 [dy/dx]. Use ▶ or ◀ to move the cursor as close as possible to $n = X = 0.6$. Press ENTER.	0.6 is not a trace value.	This is just an estimate.

We could have obtained a much better estimate than the one given above if we had found the slope of the tangent line at $X = 0.6$ rather than a value close to that number. The TI-83 allows you to do this on the home screen or graphically. Here we illustrate the graphical method.

Have $E(0.8, n)$, using X as the input variable, in Y1. Draw the tangent line to $E(0.8, n)$ at $n = 0.6$ by first drawing the graph of the function with GRAPH . With the graph on the screen, press 2nd PRGM (DRAW) 5 [Tangent(] . 6 ENTER .

Because this method gives the equation of the line tangent to the graph at the chosen input, you can see that the slope of the tangent, $\frac{dE(0.8, n)}{dn}$ at $n = 0.6$, is about -0.503 foot/mile.

9.3.3 FINDING PARTIAL RATES OF CHANGE USING CROSS-SECTIONAL MODELS

The procedures described in the last section required that you use X as the input variable to find partial rates of change. The process of replacing other variables with X can be confusing. You can avoid having to do this replacement when you use the TI-83 numerical derivative on the home screen to evaluate partial rates of change at specific input values. We illustrate with the cross-sectional function $E(e, 0.6) = -10.124e^3 + 21.347e^2 - 13.972e + 802.809$ feet above sea level that is given in Example 2 of Section 9.3 in *Calculus Concepts*. We also demonstrate this method with the cross-sectional function $E(0.8, n)$ used in the last section of this *Guide*.

Have $E(0.8, n) = -2.5n^2 + 2.497n + 799.490$, using N as the input variable, in Y1. Enter $E(e, 0.6)$, using E as the input variable, in Y2.

(Remember that because we are not using X as the input variable, we should not draw a graph or use the TABLE.)

NOTE: These functions could be entered on the home screen instead of in the Y= list. We use Y1 and Y2 in the Y= list because they are convenient locations to hold the equations.

Tell the TI-83 the point at which the derivative is to be evaluated by storing the values. Then, on the home screen, find $\frac{dE(0.8, n)}{dn}$ evaluated at $n = 0.6$ with the numerical derivative nDeriv.	`.6→N        .6`
	`.8→E        .8`
		`.6→N        .6`
		`.8→E        .8`
		`nDeriv(Y₁,N,.6)`
		`         -.503`

Find $\frac{dE(e, 0.6)}{de}$ evaluated at $e = 0.8$ to be about 0.745 foot per mile.
(Always remember to attach units of measure to the numerical values.)
	It is not necessary to again store the values for N and E unless for some reason they have been changed.	`.6`
		`.8→E        .8`
		`nDeriv(Y₁,N,.6)`
		`         -.503`
		`nDeriv(Y₂,E,.8)`
		`       .74510988`

If you want to see the line tangent to the graph of $E(e, 0.6)$ at $e = 0.8$, first change E in Y2 to X and turn off Y1. Draw the graph and use the Tangent instruction in the DRAW menu.	`Plot1 Plot2 Plot3`
	`\Y₁=-2.5N²+2.497`
	`N+799.490`
	`\Y₂=-10.124X^3+2`
	`1.347X²-13.972X+`
	`802.809`
	`\Y₃=`
	`\Y₄=`
		(graph with tangent line)
		`X=.8`
		`y=.74510988X+799.513904`

9.4 Compensating for Change

As you have just seen, your TI-83 closely estimates numerical values of partial derivatives using its nDeriv function. This technique can also be very beneficial and help you eliminate many potential calculation mistakes when you find the rate of change of one input variable with respect to another input variable (that is, the slope of the tangent line) at a point on a contour curve.

9.4.1 EVALUATING PARTIAL DERIVATIVES OF MULTIVARIABLE FUNCTIONS

The last few sections of this *Guide* indicate how to estimate and evaluate partial derivatives using cross-sectional models. The TI-83 evaluates partial derivatives calculated directly from multivariable function formulas using the same procedures. The most important thing to remember is that you must supply the name of the input variable that is changing and the values at which the partial derivative is evaluated. We illustrate using the body-mass index function that is in Example 1 of Section 9.4 of *Calculus Concepts*:

A person's body-mass index is given by $B(h, w) = \frac{0.4536w}{0.00064516h^2}$ where h is the person's height in inches and w is the person's weight in pounds. We first find B_h and B_w at a specific height and weight and then use those values in the next section of this *Guide* to find the value of the derivative $\frac{dw}{dh}$ at that particular height and weight. The person in this example is 5 feet 7 inches tall and weighs 129 pounds.

Enter B in the Y1 location of the Y= list, using the letters H and W for the input variables.
Next, store the values of H and W at the given point.
	`Plot1 Plot2 Plot3`
	`\Y₁=.4536W/(.000`
	`64516H²)`
	`\Y₂=`
	`\Y₃=`
	`\Y₄=`
	`\Y₅=`
	`\Y₆=`
		`67→H        67`
		`129→W       129`

CAUTION: The most common mistake made in using this method is forgetting to store the values of the point at which the derivative is to be evaluated. The nDeriv instruction specifies the value at only one of the inputs, and you must tell the calculator the values of all inputs.

| Find the value of $B_h$ at $h = 67$ and $w = 129$ by evaluating the expression shown to the right. The symbol $B_h$ (or $\partial B/\partial h$) tells you that $h$ is varying and $w$ is constant, so $h$ is the variable you type in the numerical derivative. Enter the value of $h$ after the letter.<br><br>Store this value in N for use in the next section of this *Guide*. | ```
nDeriv(Y₁,H,67)
           -.6031160845
Ans→N
           -.6031160845
``` |
| Find the value of $B_w$ at $h = 67$ and $w = 129$ by evaluating the expression shown to the right. The symbol $B_w$ (or $\partial B/\partial w$) tells you that $w$ is varying and $h$ is constant, so $w$ is the variable you type in the numerical derivative. Enter the value of $w$ after the letter.<br><br>Store this value in D for use in the next section of this *Guide*. | ```
nDeriv(Y₁,W,129)
             .156623169
Ans→D
             .156623169
``` |

9.4.2 FINDING THE SLOPE OF A LINE TANGENT TO A CONTOUR CURVE

We continue the previous illustration with the body-mass index function in Example 1 of Section 9.4 of *Calculus Concepts*. Part *a* of Example 1 asks for $\frac{dw}{dh}$ at the point (67, 129) on the contour curve corresponding to the person's current body-mass index. The formula is $\frac{dw}{dh} = \frac{-B_h}{B_w}$.

An easy way to remember this formula is that whatever variable is in the numerator of the derivative (in this case, w) is the same variable that appears as the changing variable in the denominator of the slope formula. This is why we stored B_w as D (for denominator) and B_h as N (for numerator). Don't forget to put a minus sign in front of the numerator.

| In the previous section, we stored $B_h$ as N and $B_w$ as D. So, $\frac{dw}{dh} = \frac{-B_h}{B_w} = -\text{N} \div \text{D}$. (Storing these values also avoids round-off error.) The rate of change is about 3.85 pounds per inch. | ```
-N/D
           3.850746274
``` |

9.4.3 COMPENSATING FOR CHANGE

When one input of a two-variable multivariable function changes by a small amount, the value of the function is no longer the same as it was before the change. The methods illustrated below show how to determine the amount by which the other input must change so that the output of the function remains at the value it was before any changes were made. We again continue the previous illustration with the body-mass index function and part *b* of Example 1 of Section 9.4 of *Calculus Concepts*.

| To estimate the change in weight needed to compensate for growths of 0.5 inch, 1 inch, and 2 inches if the person's body-mass index is to remain constant, you need to find $$\Delta w \approx \frac{dw}{dh}(\Delta h)$$ at the given values of $\Delta h$.<br><br>Again, to avoid rounding error, it is easiest to store the slope of the tangent line to some location, say T, and use the unrounded value to calculate. | ```
           3.850746274
Ans→T
           3.850746274
T*.5
           1.925373137
T*2
           7.701492549
```<br><br>These are changes in weight, so the units that should be attached are *pounds*. |

Chapter 10 Analyzing Multivariable Change: Optimization

10.2 Multivariable Optimization

As you might expect, multivariable optimization techniques that you use with your TI-83 are very similar to those that were discussed in Chapter 5. The basic difference is that the algebra required to get the expression that comes from solving a system of equations with several unknowns reduced to one equation in one unknown is sometimes difficult. However, once your equation is of that form, all the optimization procedures are basically the same as those that were discussed previously.

10.2.1 FINDING CRITICAL POINTS USING MATRICES

Critical points for a multivariable function are points at which maxima, minima, or saddle points occur. To find critical point(s) of a smooth, continuous multivariable function, we first find the partial derivatives with respect to each of the input variables and then set the partial derivatives equal to zero. This gives a *system of equations* that needs to be solved. Here we illustrate solving this system using an orderly array of numbers that is called a *matrix*.

WARNING: The matrix solution method applies only to *linear* systems of equations. That is, the system of equations should not have any variable appearing to a power higher than 1 and should not contain a product of any variables. If the system is not linear, you must use algebraic solution methods to solve the system. Using your TI-83 with the algebraic solution method is illustrated in Section 10.2.2 of this *Guide*.

Consider the function $P(h, c)$ that gives the percentage of sugar converted to olestra when the ration of peanut oil to sugar is 10 to 1. This function is given at the beginning of Section 10.2 of *Calculus Concepts*:

$$P(h, c) = -1924.3916 + 24.5402h + 25.5695c - 0.5927h^2 - 0.0860c^2 - 0.0810hc$$

The system of equations derived from the partial derivatives of P are

$$P_c = 25.5695 - 0.172c - 0.081h = 0$$
$$P_h = 24.5402 - 1.1854h - 0.081c = 0$$

We present two methods of forming the matrix used to solve the system of equations. <u>Regardless of which method you use, you MUST write the equations so that the constant terms are on the right-hand side of the equations and the coefficients of the input variables occupy the same positions on the left-hand side of each equation</u>:

$$-0.081h - 0.172c = -25.5695 \qquad [1]$$
$$-1.1854h - 0.081c = -24.5402 \qquad [2]$$

The matrix menu is accessed with MATRX. You may or may not have numbers (or the same numbers as shown below) in the column next to the matrix names A, B, etc. The *dimension* of a matrix is the number of rows and columns it contains.

Method 1: Using a Matrix of Coefficients

We choose to use A as the matrix of coefficients. Because there are 2 input variables and 2 equations, we set the dimension of A to 2 by 2; that is, A will have 2 rows and 2 columns.

Press MATRX and then ▶ ▶ to go to the EDIT menu. Press ENTER. Change the dimension of A by pressing ENTER 2 ENTER 2 ENTER.	NAMES MATH **EDIT** **1:**[A] 3×3 2:[B] 3×1 3:[C] 4:[D] 5:[E] 6:[F] 7↓[G]	MATRIX[A] 2 ×2 [**0** 0] [0 0] 1,1=0
Using equations [1] and [2] on the previous page, enter the coefficients of h in the first column and the coefficients of c in the second column of matrix A.	MATRIX[A] 2 ×2 [-.081 -.172] [**-1.1854** -.081] 2,1=-1.1854	The numbers at the bottom of the screen give the row and column that is highlighted. "2, 1" means that −1.1854 is in the 2nd row and the 1st column.

NOTE: You can move around the matrix with ENTER or by using ▶, ◀, ▲, or ▼.

Return to the home screen. We choose to use B as the constant matrix. Follow the same steps as above to set the dimension of B to 2 by 1. Enter the complete values of the constant terms.	NAMES MATH **EDIT** 1:[A] 2×2 **2:**[B] 3×1 3:[C] 4:[D] 5:[E] 6:[F] 7↓[G]	MATRIX[B] 2 ×1 [-25.57] [**-24.5402**] 2,1=-24.5402
Return to the home screen. The matrix that holds the solution to the system of equations, provided a solution exists, is obtained with MATRX 1 [A] x^{-1} MATRX 2 [B] ENTER.	[A]⁻¹[B] [[10.89448572] [143.529341]]	Carefully notice the order in which you entered the coefficients. Because the coefficient of h was input first in matrix A, the first value in the solution is the value of h. The other value is c.
We need to use these unrounded values to find the multivariable function output at the critical point. So, store this solution in another matrix, say C, with STO▶ MATRX 3 [C]. Also, enter the multivariable function P in Y1.	[A]⁻¹[B] [[10.89448572] [143.529341]] Ans→[C] [[10.89448572] [143.529341]]	Plot1 Plot2 Plot3 \Y1■-1924.3916+2 4.5402H+25.5695C -.5927H²-.0860C² -.0810HC \Y2= \Y3= \Y4=

We have found that $h \approx 10.9$ hours and $c \approx 143.5$°C. We illustrate how to recall a particular matrix element so that we can use the unrounded values of h and c to find the multivariable function output. Recall that matrix elements are referred to by the row and then the column in which they appear. Because C is a 2 by 1 matrix (2 rows and 1 column) and we set up the equations so that h was the input variable whose coefficient we entered first in matrix A, h is in the (1, 1) position and c is in the (2, 1) position.

Store the value of h into H with MATRX 3 [C] (1 , 1) STO▶ ALPHA ^ (H) ENTER. Store the value of c into C with MATRX 3 [C] (2 , 1) STO▶ ALPHA PRGM (C) ENTER. Find $P(h, c)$ with VARS ▶ [Y-VARS] 1 [Function] 1 [Y1] ENTER.	[C](1,1)→H
	    10.89448572
	[C](2,1)→C
	    143.529341
	Y1
	    44.27157195

Method 2: Using a Single Matrix

We choose to use A as the single matrix. The system of equations contains 2 input variables, a constant term in each equation, and 2 equations, so we set the dimension of A to 2 by 3; that is, A will have 2 rows and 3 columns. (Your TI-83 screens corresponding to the screens shown below may have numbers different from those indicated.)

Press MATRX and then ▶ ▶ to go to the EDIT menu. Press ENTER. Change the dimension of A by pressing ENTER 2 ENTER 3 ENTER.

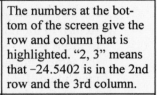

NOTE: You can move around the matrix with ENTER or by using ▶, ◀, ▲, or ▼.

Using equations [1] and [2] on page A-98, enter the coefficients of h in the first column, the coefficients of c in the second column, and the constant terms in the third column of matrix A.

The numbers at the bottom of the screen give the row and column that is highlighted. "2, 3" means that -24.5402 is in the 2nd row and the 3rd column.

Return to the home screen. The keystrokes that give the solution to the system of equations, provided that a solution exists, are MATRX ▶ [MATH] ALPHA MATRX [rref(] MATRX 1 [A] [)] ENTER. Press and hold ▶ to view all of the third column.

Carefully notice the order in which you entered the coefficients. Because the coefficient of h was input first in matrix A, the first value in the third column of the solution is the value of h. The second value in the third column is c.

We need to use these unrounded values to find the multivariable function output at the critical point. So, store this solution in another matrix, say C, with STO▶ MATRX 3 [C]. Also, enter the multivariable function P in Y1.

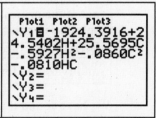

We illustrate how to recall a particular matrix element so that we can use the unrounded values of h and c to find the multivariable function output. Recall that matrix elements are referred to by the row and then the column in which they appear. Because C is a 2 by 3 matrix (2 rows and 3 column) and we set up the equations so that h was the input variable whose coefficient we entered first in matrix A, h is in the (1, 3) position and c is in the (2, 3) position.

Store the value of h into H with MATRX 3 [C] [(] 1 [,] 3 [)] STO▶ ALPHA ^ (H) ENTER. Store the value of c into C with MATRX 3 [C] [(] 2 [,] 3 [)] STO▶ ALPHA PRGM (C) ENTER. Find $P(h, c)$ with VARS ▶ [Y-VARS] 1 [Function] 1 [Y1] ENTER.

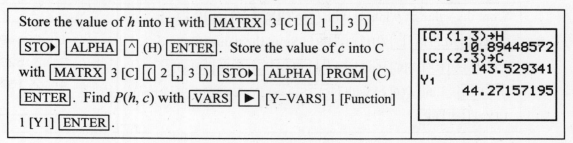

TI-83 Guide

10.2.2 FINDING CRITICAL POINTS USING ALGEBRA AND THE SOLVER We begin the process of finding critical points of a smooth, continuous multivariable function by using derivative formulas to find the partial derivative with respect to each input variable and setting these partial derivatives each equal to 0. We next use algebraic methods to obtain one equation in one unknown input. Then, you can use your calculator's solve routine to obtain the solution to that equation. This method of solution works for all types of equations. We illustrate these ideas with the function in Example 2 of Section 10.2. (Note that the partial derivative equations in Example 2 are not linear; thus, matrix solution methods can not be used.)

The total daily intake of organic matter required by a beef cow grazing the Northern Great Plains rangeland can be modeled by

$$I(s, m) = 8.61967 - 1.244s + 0.0897s^2 - 0.20988m + 0.035947m^2 + 0.214915sm$$

kg per day when the cow produces m kilograms of milk per day and s is a number between -4 and 4 indicating the size of the cow.

Find the two partial derivatives and set each of them equal to 0 to obtain these equations:

I_s: $\quad -1.244 + 0.1794s + 0.214915m = 0 \quad$ [3]

I_m: $\quad -0.20988 + 0.071894m + 0.214915s = 0 \quad$ [4]

WARNING: Everything that you do with your calculator depends on the partial derivative formulas that you find using derivative rules. Double check this work before you use any of these solution methods.

Next, solve one of the equations for one of the variables, say equation 3 for s, to obtain

$$s = \frac{1.244 - 0.214915m}{0.1794}$$

Remember that to *solve* for a quantity means that it must be by itself on one side of the equation and the other side of the equation does not contain that letter. Now, let your TI-83 work.

Clear the Y= list. Enter the expression for s in Y2. (We later enter the function I in Y1.) Be sure to enclose the numerator in parentheses. Type the left-hand side of the *other* equation (here, equation 4) in Y3. (The expression you type should be equal to zero.)	Plot1 Plot2 Plot3 \Y1= \Y2=(1.244-.2149 15M)/.1794 \Y3= \Y4= \Y5= \Y6=	Plot1 Plot2 Plot3 \Y1= \Y2=(1.244-.2149 15M)/.1794 \Y3=-.20988+.071 894M+.2149155 \Y4= \Y5=
Now, put the cursor on the letter S in the Y3 location. Replace *every* S in Y3 with Y2. What you have just done is substitute the expression for S from equation 3 (which the TI-83 knows as Y2) in equation 4! The expression in Y3 is the left-hand side of an equation that equals 0 and it contains only one variable, namely m.		Plot1 Plot2 Plot3 \Y1= \Y2=(1.244-.2149 15M)/.1794 \Y3=-.20988+.071 894M+.214915Y2 \Y4= \Y5=
The next step is to use the SOLVER to solve the equation Y3 = 0. Try several different guesses and see that, for this equation, they all result in the same solution. (See Section 1.1.7 if you need to review using the SOLVER.)	EQUATION SOLVER eqn:0=Y3	Y3=0 •M=6.8998840512... bound={-1E99,1... •left-rt=0

NOTE: You need to examine the equation in Y3 and see what type of function it represents. In this case, Y3 contains m to no power higher than one, so it is a linear equation and has only one root. If Y3 contains the variable squared, the equation is quadratic and you need to try different guesses because there could be two solutions, and so forth.

Return to the home screen and press ALPHA ÷ ENTER to check that the TI-83 knows that the answer from the SOLVER is a value of M.
 	
Next, find s by evaluating Y2. The TI-83 does not know that the output of Y2 equals S, so be sure to store this value in S.
	M
	  6.899884051
	Y2
	  -1.331597441
	Ans→S
	  -1.331597441

The last thing to do is to find the output $I(s, m)$ at the current values of s and m. To do this, enter the function I in Y1 and call up Y1 on the home screen.	Plot1 Plot2 Plot3 \Y1■8.61967-1.2445S+.08975S²-.20988M+.035947M²+.214915SM \Y2■(1.244-.214915M)/.1794 \Y3■-.20988+.071 ... Y1 8.723849776

The critical point has coordinates $s \approx -1.33$, $m \approx 6.90$ kg of milk, and $I \approx 8.72$ kg of organic matter eaten each day.

10.2.3 CLASSIFYING CRITICAL POINTS USING THE DETERMINANT TEST

Once you find one or more critical points, the next step is to classify them as points at which a maximum, a minimum, or a saddle point occurs. The Determinant Test often will give the answer. Also, because this test uses derivatives, the calculator's numerical derivative nDeriv can help.

We illustrate with the critical point that was found for Example 2. To use the Determinant Test, we need to calculate the four second partial derivatives of I and evaluate them at the critical point values of S and M.

Enter the function I in Y1, I_S in Y2 and I_m in Y3. These quantities are given on page 101 of this *Guide*.
 	
M and S should contain the unrounded values of the inputs at the critical point.
	Plot1 Plot2 Plot3
	8M+.035947M²+.214915SM
	\Y2■-1.244+.1794S+.214915M
	\Y3■-.20988+.071894M+.214915S
	\Y4=
		M
		  6.899884051
		S
		  -1.331597441

Take the derivative of Y2 = I_S with respect to s and we have I_{ss}, the partial derivative of I with respect to s and then s again. Find $I_{ss} \approx 0.1794$ at the critical point.		nDeriv(Y2,S,S)
		  .1794
		nDeriv(Y3,S,S)
		  .214915

Take the derivative of Y3 = I_m with respect to s and we have I_{ms}, the partial derivative of I with respect to m and then s. Find $I_{ms} \approx 0.2149$ at the critical point.		If you prefer, enter the value of S in the 3rd position.

Take the derivative of Y2 = I_S with respect to m and we have I_{sm}, the partial derivative of I with respect to s and then m. Find $I_{sm} \approx 0.2149$ at the critical point.		nDeriv(Y2,M,M)
		  .214915
		nDeriv(Y3,M,M)
		  .071894

Take the derivative of Y3 = I_m with respect to m and we have I_{mm}, the partial derivative of I with respect to m and then m again. Find $I_{mm} \approx 0.0719$ at the critical point.		If you prefer, enter the value of M in the 3rd position.

The second partials matrix is $\begin{bmatrix} I_{ss} & I_{sm} \\ I_{ms} & I_{mm} \end{bmatrix}$. Find the value of $D \approx (0.1794)(0.71894) - (0.214915)^2 \approx -0.033$. Because $D < 0$, the Determinant Test tells us that $(s, m, I) \approx (-1.33, 6.90, 8.72)$ is a saddle point.

```
.1794(.071894)-.
214915²
           -.0332906736
```

NOTE: The values of the second partial derivatives were not very difficult to determine without the calculator in this example. However, with a more complicated function, we strongly suggest using the above methods to provide a check on your analytic work to avoid making simple mistakes. You can also use the function in Y1 to check the derivatives in Y2 and Y3. We use some randomly chosen values of S and M to illustrate this quick check rather than use the values at the critical point, because the critical point values give 0 for Y2 and Y3.

Store different values in S and M and evaluate Y2 at these values. Then evaluate the TI-83's derivative of Y1 with respect to S at these same values.

Evaluate Y3 at the stored values of S and M. Then evaluate the TI-83's derivative of Y1 with respect to M at these same values.

```
2.53→S:8.96→M
            8.96
Y₂
         1.1355204
nDeriv(Y₁,S,2.53
)
         1.1355204
```

```
)
         1.1355204
Y₃
          .97802519
nDeriv(Y₁,M,8.96
)
          .97802519
```

10.3 Optimization Under Constraints

Optimization techniques on your calculator when a constraint is involved are the same as the ones discussed in Sections 10.2.1 and 10.2.2 except that there is one additional equation in the system of equations to be solved.

10.3.1 FINDING OPTIMAL POINTS UNDER CONSTRAINED OPTIMIZATION

You can use any of the methods indicated in Sections 10.2.1 or 10.2.2 of this *Guide* to find critical points under constrained optimization. We choose to illustrate the process using Method 2 for matrices to solve the linear system that results in the sausage cooking illustration that is at the beginning of Section 10.3 of *Calculus Concepts*. The system of equations, written with the constant terms on the right-hand side of the equations and the coefficients of the three input variables in the same positions on the left-hand side of each equation, is

$$-5.83s - \lambda = -11.78$$
$$-5.83w \qquad - \lambda = -11.69$$
$$w + s \qquad = 1$$

(The order in which the variables appear does not matter as long as it is the same in each equation. Note that we use a different order than in the text.)

When a particular variable does not appear in an equation, its coefficient is zero. Go to the MATRX [EDIT] menu, and set the dimensions of matrix A to be 3 by 4 (3 rows and 4 columns).

Enter the coefficients of w in the first column, the coefficients of s in the second column, the coefficients of λ in the third column, and the constant terms in the fourth column of matrix A. Find the solution from the home screen with [rref(] in the MATRX MATH menu.

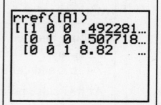

(See the discussion of Method 2 on page A-100 of this *Guide* for more detailed instructions.) The solution to the system of equations is $w \approx 0.492$, $s \approx 0.508$, and $\lambda = 8.82$.

10.3.2 CLASSIFYING OPTIMAL POINTS UNDER CONSTRAINED OPTIMIZATION

Classifying critical points when a constraint is involved is done by graphing the constraint on a contour graph or by examining close points. Your TI-83 cannot help with the contour graph classification. We illustrate the procedure used to examine close points for the function given in Example 1 of Section 10.3 of *Calculus Concepts* – the Cobb-Douglas production function $f(L, K) = 48.1L^{0.6}K^{0.4}$ subject to the constraint $g(L, K) = 8L + K = 98$.

We first find the critical point(s). Because this function does not give a linear system of equations, we use the algebraic method indicated in Section 10.2.2. The system of equations is

$$\begin{cases} 28.86L^{-0.4}K^{0.4} = 8\lambda \\ 19.24L^{0.6}K^{-0.6} = \lambda \\ 8L + K = 98 \end{cases} \Rightarrow \begin{array}{l} 28.86L^{-0.4}K^{0.4} = 8(19.24L^{0.6}K^{-0.6}) \qquad [5] \\ \\ K = 98 - 8L \qquad [6] \end{array}$$

Clear the Y= list. Enter the function *f* in Y1 and the expression for *K* in Y2. Rewrite the *other* equation (equation 5) so that it equals 0, and enter the non-zero side in Y3. Substitute* Y2 into Y3.	`Plot1 Plot2 Plot3`
	`\Y1=48.1L^.6K^.4`
	`\Y2=98-8L`
	`\Y3=28.86L^-.4K^`
	`.4-8(19.24L^.6K^`
	`-.6)`
	`\Y4=` `Plot1 Plot2 Plot3`
	`\Y1=48.1L^.6K^.4`
	`\Y2=98-8L`
	`\Y3=28.86L^-.4Y2`
	`^.4-8(19.24L^.6Y`
	`2^-.6)`
	`\Y4=`

*Remember that K = Y2. Put the cursor on the first K in Y3 and replace K by Y2. Do the same for the other K in Y3. The expression now in Y3 is the left-hand side of an equation that equals 0 and contains only one variable, namely L. (We are not sure how many answers there are to this equation.)

Use the SOLVER to solve the equation Y3 = 0. Try several different guesses and see that they all result in the same solution. On the home screen call up Y2 to find K. (Store this value in K.) Y1 gives the value of *f* at this point.	`Y3=0`
	`■L=7.35`
	`  bound={-1E99,1…`
	`■left-rt=0` `Y2          7.35`
	`            39.2`
	`Ans→K`
	`            39.2`
	`Y1`
	`     690.6084798`

We now need to test close points to see if this output value of *f* is maximum or minimum. You need to remember that whatever close points you choose, they must be near the critical point and *they must be on the constraint g*.

WARNING: Do not round during this procedure. Rounding of intermediate calculations and/or inputs can give a false result when the close point is very near to the optimal point.

Choose a value of L that is less than L = 7.35, say 7.3. Find the value of *K* so that 8L + K = 98. Remember that K = Y2, so store 7.3 in L and call up Y2. Find *f* at L = 7.3, K = 39.6.
 	
At this close point, the value of *f* is less than the value of *f* at the optimal point.
	`7.3→L:Y2`
	`            39.6`
	`Ans→K`
	`            39.6`
	`Y1`
	`     690.5845641`

Choose another value of L, this time one that is more than L = 7.35, say 7.4. Find the value of *K* so that 8L + K = 98 by storing 7.4 in L and calling up Y2. Find *f* at L = 7.4, K = 38.8.
 	
At this close point, the value of *f* is less than the value of *f* at the critical point. Thus, (7.35, 39.2, 690) is a maximum point.
	`7.4→L:Y2`
	`            38.8`
	`Ans→K`
	`            38.8`
	`Y1`
	`     690.5844554`

Chapter 11 Dynamics of Change: Differential Equations and Proportionality

11.3 Numerically Estimating by Using Differential Equations: Euler's Method

Many of the differential equations we encounter have solutions that can be found by determining an antiderivative of a given rate-of-change function. Thus, many of the techniques that we learned using the TI-83's numerical integration function apply to this chapter. (See Chapters 6 and 7 of this *Guide*.)

11.3.1 EULER'S METHOD FOR A DIFFERENTIAL EQUATION WITH ONE INPUT VARIABLE You may encounter a differential equation that cannot be solved by standard methods and you may need to draw an accumulation graph for a differential equation without first finding an antiderivative. In either of these cases, numerically estimating a solution using Euler's method is helpful. This method relies on the use of the derivative of a function to approximate the change in that function. Recall from Section 5.1 of *Calculus Concepts* that the approximate change in f at a point is the rate of change of f at that point times a small change in x. That is,

$$f(x + h) - f(x) \approx f'(x) \cdot h \text{ where } h \text{ represents the small change in } x.$$

Now, if we let $b = x + h$ and $x = a$, the above expression becomes

$$f(b) - f(a) \approx f'(a) \cdot (b - a) \quad \text{or} \quad f(b) \approx f(a) + (b - a) \cdot f'(a)$$

The starting values for the coordinates of the point (a, b) will be given to you and are often called the *initial condition*. The next step is to repeatedly apply the formula given above to use the slope of the tangent line at $x = a$ to approximate the change in the function between the inputs a and b. When h, the distance between a and b, is fairly small, Euler's method will often give close numerical estimates of points on the solution to the differential equation containing $f'(x)$.

WARNING: Be wary of the fact that there is some error involved in each step of the Euler approximation process that is compounded when each result is used to obtain the next result.

We illustrate Euler's method for a differential equation containing one input variable with the differential equation in Example 1 of Section 11.3. This equation gives the rate of change of the total sales of a computer product t years after the product was introduced:

$$\frac{dS}{dt} = \frac{6.544}{\ln(t + 1.2)} \text{ billion dollars per year}$$

Because Euler's method involves a repetitive process, a program that performs the calculations used to find the approximate change in the function can save you time and eliminate computational errors and some error in rounding.

Before using this program, you must have the differential equation in the Y1 location of the Y= list with X as the input variable. Access program Euler with PRGM .

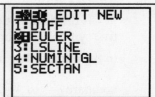

Note that your program list may not be the same as the one shown above. The code for program EULER is listed in the *TI-83 Program Appendix*.

| Run the program. Each time the program stops for input or for you to view a result, press ENTER to continue.<br><br>We choose to use 16 steps. Enter this value. The interval is 4 years, so enter the step size = $\frac{\text{length of interval}}{\text{number of steps}} = \frac{4}{16} = 0.25$. | ```
HAVE DY/DX IN Y1
NUMBER OF STEPS=
16
STEP SIZE= 0.25■
``` | |
| The initial condition is given as the point (1, 53.2). Enter these values when prompted for them. | ```
INITIAL INPUT=
1
INITIAL OUTPUT=
53.2
INPUT,OUTPUT IS
           1.25
     55.27493782
``` | The first application of the formula gives an estimate for the value of the total sales at $x = 1.25$:<br><br>$S(1.25) \approx 55.275$ |
| Press ENTER several more times to obtain more estimates for total sales. Record the input values and output estimates on paper as the program displays them. | ```
     57.10065132
INPUT,OUTPUT IS
           1.75
     58.74776643
INPUT,OUTPUT IS
           2
     60.26005352
``` | Continue pressing ENTER to obtain more estimates of points on the total sales function $S$. |
| When 16 steps have been completed (that is, after the input reaches 5), the program draws a graph of the points *(input, output estimate)* connected with straight line segments. This solution graph is an estimate of the graph of the differential equation solution. | ```
     71.70209612
INPUT,OUTPUT IS
           4.75
     72.64207417
INPUT,OUTPUT IS
           5
     73.55942757
``` | This is an estimate of the graph of the function $S(t)$. |
| Press Y=, turn Plot3 off, turn Y1 on, and press ZOOM ▲ [Zoomfit] to draw the graph of the differential equation. | ```
Plot1 Plot2 Plot3
\Y1■6.544/ln(X+1
.2)
\Y2=
\Y3=
\Y4=
\Y5=
\Y6=
``` | This is the slope graph – the graph of $S'(t)$. |

11.3.2 EULER'S METHOD FOR A DIFFERENTIAL EQUATION WITH TWO INPUT VARIABLES

Program EULER can be used when the differential equation is a function of x and y with $y = f(x)$. Follow the same process that is illustrated in the previous section of this *Guide*, but enter $\frac{dy}{dx}$ in Y1 using the letters x and y as they are written in the given equation.

If the differential equation is written in terms of variables other than x and y, let the derivative symbol be your guide as to which variable corresponds to the input and which corresponds to the output. For instance, if the rate of change of a quantity is given by $\frac{dP}{dn} = 1.346P(1 - n^2)$, compare $\frac{dy}{dx}$ to $\frac{dP}{dn}$, entering in Y1 the expression 1.346Y(1 − X^2). Use ALPHA 1 (Y) to type Y.
```

TI-83 Guide

The differential equation may be given in terms of y only. For instance, if $\frac{dy}{dx} = k(30 - y)$ where k is a constant, enter Y1 = K(30 – Y). Of course, you need to store a value for k or substitute a value for k in the differential equation before using program EULER. It is always better to store the exact value for a constant instead of using a rounded value.

We illustrate using Euler's method with two input variables by using the equation in Example 2 of Section 11.3 in *Calculus Concepts*.

Enter $\frac{dy}{dx} = 5.9x - 3.2y$ in Y1. Run program EULER. We choose to use 10 steps over an interval of length 2; hence, the step size is 0.2. The initial condition is $y(10) = 50$.	Plot1 Plot2 Plot3 \Y1■5.9X-3.2Y \Y2= \Y3= \Y4= \Y5= \Y6= \Y7=	NUMBER OF STEPS= 10 STEP SIZE= .2 INITIAL INPUT= 10 INITIAL OUTPUT= 50■

The first estimate for a point on the solution is $y = 29.8$ when $x = 10.2$.
Continue until 10 steps are completed; that is, until the input is 12.
	INITIAL INPUT=
	10
	INITIAL OUTPUT=
	50
	INPUT,OUTPUT IS
	  10.2
	  29.8
		20.8203948
		INPUT,OUTPUT IS
		  11.8
		  21.18334213
		INPUT,OUTPUT IS
		  12
		  21.55000317

The estimate of $y(12)$ is 21.55. Press ENTER to draw a graph of the Euler estimates. (Note that we cannot graph the differential equation on the TI-83 as we do when there is only one input variable.)		

Appendix: Trigonometry Basics

Before you begin working through the ideas in this *Appendix*, be certain that the MODE screen in the basic setup has Radian/Degree set to Radian measure.

A.1 FINDING OUTPUTS OF TRIG FUNCTIONS WITH DEGREE INPUTS
It is very important that you have the correct mode set when evaluating trigonometric function outputs. Unless you see the degree symbol in a problem, the mode setting should be Radian. We show two different methods of evaluating trig functions with degree inputs, using the illustration as given in the discussion following Figure A9 in the *Trigonometry Appendix*.

To find cos 25° with MODE set to Radian, press COS 25 and then tell the TI-83 that the angle is measured in degrees by pressing 2nd MATRX (ANGLE) 1 (°)) ENTER.		cos(25°) .906307787
We now find sin 25° using a different method: change the MODE setting to Degree, return to the home screen, type SIN 25), and press ENTER.	Normal Sci Eng Float 0123456789 Radian **Degree** **Func** Par Pol Seq **Connected** Dot **Sequential** Simul **Real** a+bi re^θi **Full** Horiz G-T	sin(25) .4226182617

WARNING: Do not forget to change the MODE setting back to Radian if you use this second method. If not, you will obtain an incorrect value the next time you evaluate a trig function for an angle measured in radians or when drawing a graph. For this reason, we recommend using the first method.

A.2 CONVERTING ANGLES FROM RADIANS TO DEGREES Even though the formula for this conversion is simple, the TI-83 converts angles from radians to degrees quite easily. We illustrate using the angle in part *a* of Example 7 in the *Trigonometry Appendix*.

As shown in Example 7, convert $7\pi/6$ radians to degrees with multiplication by the proper factor. Let the units be your guide and carefully use parentheses: $\left(\frac{7\pi}{6} \text{ radians}\right)\left(\frac{180°}{\pi \text{ radians}}\right) = 210°$

Note that the MODE setting does not matter because you are not using the ANGLE menu or any of the trig function keys.

To have the TI-83 do the conversion, set MODE to Degree, type $7\pi/6$, and press [2nd] [MATRX] (ANGLE) 3 (ʳ). You must enclose $7\pi/6$ in parentheses or the TI-83 assumes only the 6 is measured in radians.

NOTE: When MODE is set to Degree, the TI-83 assumes all angles (inputs and outputs) are measured in degrees. The radian setting (ʳ) in the ANGLE menu x overrides the MODE setting.

A.3 CONVERTING ANGLES FROM DEGREES TO RADIANS We illustrate converting an angle from degree measure to radian measure using the angle in part *b* of Example 7 in the *Trigonometry Appendix*. As indicated in the last section, you can convert using the fact that π radians = 180 degrees or let the TI-83 do the conversion.

As shown in Example 7, convert 135 degrees with multiplication by the proper factor. Let the units be your guide and carefully use parentheses: $\left(135°\right)\left(\frac{\pi \text{ radians}}{180°}\right) = \frac{3\pi}{4} \approx 2.36$ radians

Note that the MODE setting does not matter because you are not using the ANGLE menu or any of the trig function keys.

(See page A-2 for the keystrokes needed to access the fraction conversion key.)

To have the TI-83 do the conversion, set MODE to Radian, type 135, and press [2nd] [MATRX] (ANGLE) 1 (°). Press [ENTER].

NOTE: When MODE is set to Radian, the TI-83 assumes all angles (inputs and outputs) are measured in radians. The degree choice (°) in the ANGLE menu overrides the MODE setting.

TroubleShooting the TI-83

❏ CALCULATOR BATTERIES

- *Will I lose all my programs when I change batteries?*

There is no guarantee that changing batteries will not reset your calculator's memory. However, a method that seems to work well is to make sure your calculator is turned off, take the old batteries out one at a time, replacing each with a new battery before removing the next. Be sure you have the direction of the batteries correct (+ and – alternating).

- *What is the backup battery?*

The backup battery is a round battery above the AAA batteries and under a piece of plastic that is fastened to the calculator with a small screw.

- *How long will my batteries last?*

How long your batteries last depends on how much you use your calculator. The four AAA batteries in the calculator usually last for about a year with normal use in a calculus course. The backup battery can last anywhere from 2 to 5 years.

- *When do I need to replace the batteries?*

You need to replace the AAA batteries if your screen is not dark enough and 9 is the value shown in the upper right hand column when you press **2nd**, release that key, and then press and hold the up arrow. (See the second bulleted item under the *General* category.)

- *I've replaced the four AAA batteries and my calculator still won't come on. What next?*

Sometimes when you replace batteries, the calculator resets to its lightest screen setting. Darken the display according to the directions given in the second bulleted item under the GENERAL category. If you have replaced the AAA batteries, you have darkened the display, and your calculator still will not turn on, it is likely that the backup battery needs replacing.

❏ DELETING PROGRAMS OR OTHER INFORMATION FROM THE TI-83

- *How do I delete information from my calculator?*

Press **2nd + [MEM] 2 [Delete]**. Next press the number corresponding to what you wish to delete. For instance, to delete a program, press **7 [Prgm]**. Use the down arrow to move the cursor next to the information you want to delete and press **ENTER**. *Be very careful – there is no "undo" command.*

- *The calculator keeps giving me a memory error when I try to enter data or transfer or run a program. What can I do?*

Press **2nd + [MEM] 1 [Check RAM]**. If you have 5000 or less in the MEM FREE location, you need to delete some of the calculator's contents. Pictures (PIC) and graph data bases (GDB) take up a lot of room and can be deleted by following the directions in the answer directly above. Game programs also take up a lot of room and create some of the pictures and data bases that may be stored in your calculator.

❏ GENERAL INFORMATION

- *Why do my calculator screens not look like the ones in this Guide?*

Make sure your calculator settings are as specified in the *Setup Instructions* on page *A*-1.

- *How do I make the display screen darker or lighter?*

To make the display darker, press **2nd**, release that key, and then press and hold the up arrow until the display is dark enough. To make the display lighter, press **2nd**, release that key, and then press and hold the down arrow until the display is light enough. You should see a number in the upper right corner of the display screen that varies between 0 (light) to 9 (dark.)

- *The calculator doesn't recognize what I have entered in Y1 as the function.*

Anytime you refer to a function in the Y= list, you must type the function name using **VARS right arrow [Y-VARS] 1 [Function]** followed by the number of the desired function location. The TI-83 does not recognize **ALPHA 1 (Y)** as the name of a function in the Y= list.

- *I have either lost or never got the owner's manual for the TI-83. What can I do?*

The *TI-83 Guidebook* is available at the TI web site with address **www.TI.com/calculators**.

❑ GRAPHING

- *Why don't I see a graph when I press the correct keys?*

Make sure that you have the function entered in the graphing (Y=) list, using X for the input variable. (See Sections 1.1.1 – 1.1.2.) Also check that your function is turned on. (See Section 1.3.5.) If the function is turned on but you still cannot see the graph, check your window settings – maybe the function did graph, but it graphed outside your window.

- *What do I do if a get a strange-looking graph or no graph instead of a scatter plot of data when I press* **ZOOM 9 [ZoomStat]**?

The scatter plot setup has somehow been changed, is not correctly set, or is not turned on. Refer to Section 1.5.5 of this *Guide* for instructions.

- *Do I have to clear the function location in the Y= list before pasting in another function when finding an equation of best fit?*

Most of the time, it is not necessary to first clear any previously-entered function from the chosen location of the Y= list. However, if you receive an error message when finding the equation, clear the desired function location and press **2nd ENTER (ENTRY) ENTER** to recall and repeat the regression instruction. If you still obtain an error message, reset the statistical setup as described on the first page of this *Guide*.

- *While trying to draw the graph of an equation that I entered in the Y= list, I get a STAT-PLOT error. What do I do?*

The TI-83 will sometimes not draw the graph of a function when one of the scatter plots is turned on. To correct this problem, choose **1: [Quit]** and press **2nd Y= (STAT PLOT) 4 [Plots Off] ENTER**. The graph should now draw with out any problems.

- *What causes the error message* ERR: WINDOW RANGE *and how do I eliminate this error?*

This error message is usually caused when Ymin = Ymax in the WINDOW settings. The code in a program may be such that the height of the window (Ymin – Ymax) is set to zero. If you attempt to draw a scatter plot when the output is constant, this error results because the TI-83 uses the output data to set the range for the WINDOW. To correct this error, manually set Ymin and Ymax and press **GRAPH** to draw the graph.

❑ LISTS

- *I don't have all the lists when I press* **STAT ENTER**. *(For instance, what do I do if list* L1 *is missing?)*

Make sure your calculator settings are as specified in Figure 2 using the second bulleted item in the list describing the Setup Instructions on the first page of this *Guide*.

❑ PROGRAMS

- *Where can I get the programs that are used in this Guide?*

The programs can be transferred to your calculator from another student or your teacher's calculator and a link chord. If no one has the programs, they can be downloaded from the web site for this text accessible through **college.hmco.com**.

A TI Graph-Link™ cable and the linking software for your particular calculator are needed to transfer the programs from the Internet to a computer. A special cable is needed to transfer the programs from the computer to your calculator. The linking software is free at the TI web site with address **www.ti.com/calculators**. If you cannot find a cable, ask at the learning center for your school, check with your instructor, or search the TI web site for instructions on how to purchase one.

- *Why won't a program run, or why do I get an error when I try to run a program?*

You may have unknowingly deleted or altered one or more lines of the program code. The easiest method of fixing the problem is to use the link chord to re-transfer the entire program into the calculator. If this is not possible, you can find the correct program code in the *TI-83 Program Appendix* at the *Calculus Concepts* Website. Press **PRGM right arrow [EDIT]** and **down arrow** until the name of the corrupt program is highlighted. Press **ENTER**. Compare the code in your calculator with what is printed in the *TI-83 Program Appendix* until you find the error and correct it by retyping the correct code. Consult the *TI-83 Guidebook* that came with your calculator for the location of the symbols in the program code.

- *Program NUMINTGL keeps giving an error when I try to run it. What can I do?*

Delete all functions from the Y= list (except the function in Y1) before using the program named NUMINTGL. This program uses every memory location in the calculator. If you receive an error while running this program, you may have a picture, another program, or something else stored as a single-letter name. For instance, if program NUMINTGL is trying to store a number in memory location T and you have a program called T, the calculator stops and does not give an error message. Delete or rename any programs, pictures, or whatever you have called by a single-letter name before continuing. (See DELETING PROGRAMS OR OTHER INFORMATION FROM THE TI-83.)

Part B

Guide for Texas Instruments TI-86 Graphing Calculator

This *Guide* is designed to offer step-by-step instruction for using your TI-86 graphing calculator with the text *Calculus Concepts: An Informal Approach to the Mathematics of Change*. A technology icon next to a particular example or discussion in the text directs you to a specific portion of this *Guide*. You should also utilize the table of contents in this *Guide* to find specific topics on which you need instruction.

Setup Instructions

Before you begin, check the TI-86 setup and be sure the settings described below are chosen. Whenever you use this *Guide*, we assume (unless instructed otherwise) that your calculator settings are as shown in Figures 1, 2, and 3.

- Press [2nd] [MORE] (MODE) and choose the settings shown in Figure 1 for the basic setup.

- Check the window format by pressing [GRAPH] [MORE] [F3] (FORMT) and choose the settings shown in Figure 2.

  - If you do not have the darkened choices shown in Figure 1 and Figure 2, use the arrow keys to move the blinking cursor over the setting you want to choose and press [ENTER].

  - Return to the home screen with [EXIT] or [2nd] [EXIT] (QUIT). Note that [EXIT] [EXIT] clears the menus from the bottom of the screen.

- Specify the statistical setup as shown in Figure 3 by pressing [2nd] [−] (LIST) [F5] [OPS] [MORE] [MORE] [MORE] [F3] [SetLE] [ALPHA] 7 (L) 2 [,] [ALPHA] 7 (L) 2 [,] [ALPHA] 7 (L) 3 [,] [ALPHA] 7 (L) 4 [,] [ALPHA] 7 (L) 5 [,] [ENTER]. You need this setup for some of the programs referred to in this *Guide* to execute properly.

TI-86 Basic Setup

Figure 1

TI-86 Window Setup

Figure 2

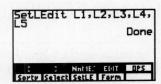

TI-86 Statistical Setup

Figure 3

Basic Operation

You should be familiar with the basic operation of your calculator. With your calculator in hand, go through each of the following.

1. **CALCULATING** You can type in lengthy expressions; just make sure that you use parentheses when you are not sure of the calculator's order of operations. Always enclose in parentheses any numerators and denominators of fractions and powers that consist of more than one term.

Evaluate $\dfrac{1}{4*15+\dfrac{895}{7}}$. Enclose the denominator in parentheses so that the addition is performed before the division into 1. It is not necessary to use parentheses around the fraction 895/7.

Evaluate $\dfrac{(-3)^4 - 5}{8 + 1.456} \approx 8.037$. Use $\boxed{(-)}$ for the negative symbol and $\boxed{-}$ for the subtraction sign. To clear the home screen, press $\boxed{\text{CLEAR}}$.

NOTE: The numerator and denominator must be enclosed in parentheses and $-3^4 \neq (-3)^4$.

Now, evaluate $e^3 * 0.027$ and $e^{3*0.027}$. Type e^ with $\boxed{\text{2nd}}$ $\boxed{\text{LN}}$ (e^x). The TI-86 will assume you mean $e^3 * 0.027$ unless you use parentheses around the two values in the exponent to indicate $e^{3*0.027}$.

2. **USING THE ANS MEMORY** Instead of again typing an expression that was just evaluated, use the answer memory by pressing $\boxed{\text{2nd}}$ $\boxed{(-)}$ (ANS).

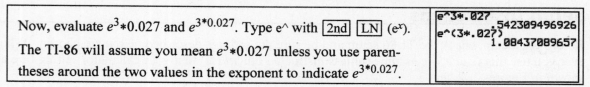

Calculate $\left(\dfrac{1}{4*15+\dfrac{895}{7}}\right)^{-1}$ using this nice shortcut.

Type Ans^{-1} by pressing $\boxed{\text{2nd}}$ $\boxed{(-)}$ (ANS) $\boxed{\text{2nd}}$ $\boxed{\text{EE}}$ (x^{-1}).

3. **ANSWER DISPLAY** When the denominator of a fraction has no more than three digits, your calculator can provide the answer in the form of a fraction. When an answer is very large or very small, the calculator displays the result in scientific notation.

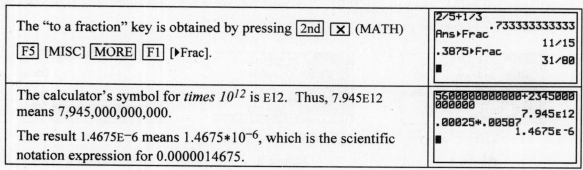

The "to a fraction" key is obtained by pressing $\boxed{\text{2nd}}$ $\boxed{\times}$ (MATH) $\boxed{\text{F5}}$ [MISC] $\boxed{\text{MORE}}$ $\boxed{\text{F1}}$ [▸Frac].

The calculator's symbol for *times* 10^{12} is E12. Thus, 7.945E12 means 7,945,000,000,000.

The result 1.4675E−6 means $1.4675 * 10^{-6}$, which is the scientific notation expression for 0.0000014675.

4. **STORING VALUES** It may be beneficial to store numbers or expressions for later recall. To store a number, type it, press STO▶ (note that the cursor changes to the alphabetic cursor **A**), press the key corresponding to the capital letter(s) naming the storage location, and then press ENTER. To join several short commands together, use 2nd . (:) between the statements. Note that when you join statements with a colon, only the value of the last statement is shown.

WARNING: The STO▶ key locks the upper-case alphabetic cursor and ALPHA unlocks it. Always look at the screen when you are typing to be certain that you are not entering numbers when you intend to type letters and vice-versa.

Store 5 in A and 3 in B, and then calculate $4A - 2B$. (Press ALPHA to return to the regular cursor.) To recall a value, press ALPHA, type the letter in which the value is stored, and then press ENTER.	5→A:3→B 3 4A−2B 14 A 5

- Storage location names on the TI-86 can be from one to eight characters long, but they must begin with a letter. You cannot name what you are storing with the exact name the TI-86 already uses for a built-in variable (such as LOG).

- Whatever you store in a particular memory location stays there until it is replaced by something else either by you or by executing a program containing that name.

NOTE: The TI-86 allows you to enter upper and lower case letters, and it distinguishes between them. For instance, VOL, Vol, VOl, vol, voL, and so forth are all different names to the TI-86. To type a lower case letter, press 2nd ALPHA before pressing a letter key (note that the cursor changes to **a**). If you cannot remember which combination of upper- and lower-case letters you used for a name, press 2nd CUSTOM [CATLG-VARS] F2 [ALL] and then press the key of the first letter of the name. Use ▼ to move the cursor next to the name and then press ENTER.

5. **ERROR MESSAGES** When your input is incorrect, the TI-86 displays an error message.

If you have more than one command on a line without the commands separated by a colon (:), an error message results when you press ENTER.	2→A x+2→B	ERROR 07 SYNTAX GOTO QUIT

- When you get an error message, press F1 [Goto] to position the cursor to where the error occurred so that you can correct the mistake or choose F5 [Quit] to begin a new line on the home screen. When you are executing a program, you should always choose the 1: Quit option upon receiving an error message. Choosing 2: Goto will call up the program code, and you may inadvertently change the program so that it will not properly execute.

A common mistake is using the negative symbol (−) instead of the subtraction sign − or vice-versa. The TI-86 does *not* give an error message, but a wrong answer results. The negative sign is shorter and raised slightly more than the subtraction sign.	2*−3 −6 2−3 −6 2−3 −1

Chapter 1 Ingredients of Change: Functions and Linear Models

1.1 Models, Functions, and Graphs

There are many uses for a function that is entered in the calculator's graphing list. Graphing the function in an appropriate viewing window is one of these. Because you must enter all functions on one line (that is, you cannot write fractions and exponents the same way you do on paper), it is very important to have a good understanding of the calculator's order of operations and to use parentheses whenever they are needed.

1.1.1 ENTERING AN EQUATION IN THE GRAPHING LIST The graphing list contains space for 99 equations (if memory is available), and the output variables are called by the names y1, y2, ..., y99. To graph an equation, enter in the graphing list. You must use x as the input variable if you intend to draw the graph of the equation or use the TI-86 table. We illustrate graphing using the equation in Example 4 of Section 1.1: $v(t) = 3.622(1.093^t)$.

Press GRAPH F1 [y(x)=] to access the graphing list. If there are any previously entered equations that you will no longer use, delete them from the graphing list with F4 [DELf].	To clear, but not delete, an equation in the y(x)= list, position the cursor on the line with the equation and press CLEAR.
For convenience, we use the first, or y1, location in the list. We intend to graph this equation, so the input variable must be called x, not t. Enter the right-hand side of the equation, $3.622(1.093^x)$, with 3 . 6 2 2 (1 . 0 9 3 ^ x-VAR) You should use the x-VAR key for x, not the times sign key, × , nor the capital letter X obtained with ALPHA + (X).	Plot1 Plot2 Plot3 \y1■3.622(1.093^x)

CAUTION: Plot1, Plot2, and Plot3 at the top of the y(x)= list should *not* be darkened when you are graphing an equation and not graphing data points. If any of these is darkened, use ▲ until you are on the darkened plot name. Press ENTER to make the name(s) not dark. If you do not do this, you may receive a STAT PLOT error message.

1.1.2 DRAWING A GRAPH As is the case with most applied problems in *Calculus Concepts*, the problem description indicates the valid input interval. Consider Example 4 of Section 1.1:

> The value of a piece of property between 1980 and 2000 is given by $v(t) = 3.622(1.093^t)$ thousand dollars where *t* is the number of years since the end of 1980.

The input interval is 1980 (*t* = 0) to 2000 (*t* = 20). Before drawing the graph of *v* on this interval, enter the *v*(*t*) equation in the y(x)= list using *x* as the input variable. (See Section 1.1.1 of this *Guide*.) We now draw the graph of the function *v* for *x* between 0 and 20.

Press GRAPH F2 [WIND] to set the view for the graph. Enter 0 for xMin and 20 for xMax. (For 10 tick marks between 0 and 20, enter 2 for Xscl. If you want 20 tick marks, enter 1 for xScl, etc. xScl does not affect the shape of the graph.	WINDOW xMin=0 xMax=20 xScl=2 yMin=-6 yMax=6 ↓yScl=1

xMin and xMax are, respectively, the settings of the left and right edges of the viewing screen, and yMin and yMax are, respectively, the settings for the lower and upper edges of the viewing screen. xScl and yScl set the spacing between the tick marks on the x- and y-axes. (Leave xRes set to 1 for all applications in this *Guide*.) We set the values for the output in the next set of instructions.)

To have the TI-86 determine the output view, press [F3] [ZOOM] [MORE] [F1] [ZFIT] [ENTER].

Note that any vertical line drawn on this graph intersects it in only one point, so the graph does represent a function. (Recall that [CLEAR] removes the menu from the bottom of the screen if you want to see the entire graphics screen.)

Press [GRAPH] [F2] [WIND] to see the view set by ZFIT. The view has $0 \leq x \leq 20$ and $3.622 \leq y \leq 21.446...$. (Note that ZFIT did not change the x-values you manually set.)

We just saw how to have the TI-86 set the view for the output variable. Whenever you draw a graph, you can also manually set or change the view for the output variable.

1.1.3 MANUALLY CHANGING THE VIEW OF A GRAPH

If for some reason you do not have an acceptable view of a graph or if you do not see a graph, change the view for the output variable with one of the ZOOM options or manually set the view for the window until you see a good graph. (We later discuss other ZOOM options.) We continue using the function v in Example 4 of Section 1.1, but we assume here that you have not yet drawn the graph of v.

Press [GRAPH] [F2] [WIND], enter 0 for xMin and 20 for xMax, and (assuming we do not know what to use for the vertical view), enter some arbitrary values for yMin and yMax. (The graph to the right was drawn with yMin = −5 and yMax = 5). Press [F5] [GRAPH].

Evaluating Outputs on the Graphics Screen: Press [F4] [TRACE]*. Recall we are given that the input is between 0 and 20. If you now type the number that you want to substitute in the function whose graph is drawn, say 0, you see the screen to the right. A 1 appears at the top of the screen because the equation of the function whose graph you are drawing is in y1.

Press [ENTER] and the input value is substituted in the function.

The input and output values are shown at the bottom of the screen. (This method works even if you do not see any of the graph on the screen.)

Substitute the right endpoint of the input interval into the function by pressing 20 [ENTER].

We see that two points on this function are approximately (0, 3.622) and (20, 21.446).

Press [GRAPH] [F2] [WIND], enter 3.5 for yMin and 22 for yMax, and press [F5] [GRAPH]. If the graph you obtain is not a good view of the function, repeat the above process using *x*-values in between the two endpoints in order to see if the output range should be extended in either direction. (Note that the choice of the values 3.5 and 22 was arbitrary. Any values close to the outputs in the points you find are also acceptable.)

*Instead of using TRACE, you could use the TI-86 TABLE or evaluate the function at 0 and 20 on the home screen to find the range of values for the output value. We later discuss using these features.

1.1.4 TRACING TO ESTIMATE OUTPUTS

You can display the coordinates of certain points on the graph by tracing. Unlike the substitution feature of TRACE that was discussed on the previous page, the *x*-values that you see when tracing the graph depend on the horizontal view that you choose. The output values that are displayed at the bottom of the screen are calculated by substituting the *x*-values into the equation that is being graphed. We again assume that you have the function $v(x) = 3.622(1.093^x)$ entered in the y1 location of the y(x)= list.

With the graph on the screen, press [F4] [TRACE], press and hold [▶] to move the trace cursor to the right, and press and hold [◀] to move the trace cursor to the left.

- Again note that the number corresponding to the location of the equation (in the y(x)= list) that you are tracing appears in the top right corner of the graphics screen.

Trace past one edge of the screen and notice that even though you cannot see the trace cursor, the *x*- and *y*-values of points on the line are still displayed at the bottom of the screen. Also note that the graph scrolls to the left or right as you move the cursor past the edge of the current viewing screen.

Use either [▶] or [◀] to move the cursor near $x = 15$. We estimate that *y* is *approximately* 13.8 when *x* is *about* 15.

It is important to realize that trace outputs should *never* be given as answers to a problem unless the displayed *x*-value is *identically* the same as the value of the input variable.

1.1.5 EVALUATING OUTPUTS ON THE HOME SCREEN

The input values used in the evaluation process are *actual* values, not estimated values such as those that are generally obtained by tracing near a certain value. Actual values are also obtained when you evaluate outputs from the graphing screen using the process that was discussed in Section 1.1.3 of this *Guide*.

We again consider the function $v(t) = 3.622(1.093^t)$ that is in Example 4 of Section 1.1.

Using *x* as the input variable, enter y1 = 3.622(1.093)^ x. Return to the home screen by pressing [2nd] [EXIT] (QUIT). Substitute 15 into the function with [2nd] [ALPHA] (alpha) 0 (Y) 1 [(] 15 [)]. Find the value by pressing [ENTER].

TI-86 Guide

NOTE: You do not have to have the *closing* parentheses on the right if nothing else follows it. To choose another graphing list location, say y2, just type the number corresponding to that function's location, 2, following the lower-case y.

WARNING: You must use a lower case, not upper case, y in order for the TI-86 to recognize the function in the graphing list. You must also use a lower-case x for the *x*-variable.

To now evaluate the function at other inputs, first recall the previous entry with [2nd] [ENTER] (ENTRY). Then edit the expression to the new value.

| For instance, press [2nd] [ENTER] (ENTRY), change 15 to 20 by pressing [◄] [◄] [◄] and typing 20, and then press [ENTER] to evaluate the function at *x* = 20. Evaluate y1 at *x* = 0 by recalling the previous entry with [2nd] [ENTER] (ENTRY), change 20 to 0 with [◄] [◄] [◄] [DEL], and then press [ENTER]. | y1(20)
        21.4462658366
y1(0)
        3.622 |

1.1.6 EVALUATING OUTPUTS USING THE TABLE

Function outputs can be determined by evaluating on the graphics screen, as discussed in Section 1.1.3, or by evaluating on the home screen as discussed in Section 1.1.5 of this *Guide*. You can also evaluate functions using the TI-86 TABLE. When you use the table, you can either enter specific input values and find the outputs or generate a list of input and output values in which the inputs begin with a value called TblStart and differ by a value called ΔTbl.

Let's use the TABLE to evaluate the function $v(t) = 3.622(1.093^t)$ at $t = 15$. Even though you can use any of the function locations, we again choose to use y1. Press [GRAPH] [F1] [y(x)=], clear the function locations, and enter 3.622(1 .093)^x in location y1 of the y(x)= list.

| After entering the function *v* in y1, choose the TABLE SETUP menu by pressing [TABLE] [F2] [TBLST]. To generate a list of values beginning with 13 such that the table values differ by 1, enter 13 in the TblStart location and 1 in the ΔTbl location. Then choose AUTO in the Indpnt: location by having the cursor on that word and pressing [ENTER]. | TABLE SETUP
TblStart=13
ΔTbl=1
Indpnt: **Auto** Ask

[TABLE] |
| Press [F1] [TABLE], and observe the list of input and output values. Notice that you can scroll through the table with [▼], [▲], [◄], and/or [►]. | x   y1
13   11.50828
14   12.57855
15   **13.74836**
16   15.02696
17   16.42446
18   17.95194
y1=13.748359355277
[TBLST][SELCT][x][y] |

- The table values may be rounded in the table display. You can see more of the output by highlighting a particular value and viewing the bottom of the screen.

| Return to the table set-up screen with [F1] [TBLST]. To compute specific outputs rather than a list of values, choose ASK in the Indpnt: location. Press [ENTER]. (Note that when using ASK, the settings for TblStart and ΔTbl do not matter.) | TABLE SETUP
TblStart=13
ΔTbl=1
Indpnt: Auto **Ask**

[TABLE] |

Press [F1] [TABLE], type in the *x*-value(s) at which the function is to be evaluated, and press [ENTER]. Unwanted entries or values from a previous problem can be cleared with [DEL].

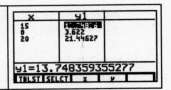

NOTE: If you are interested in evaluating a function at inputs that are not evenly spaced and/or you only need a few outputs, you should use the ASK feature of the table instead of AUTO.

1.1.7 FINDING INPUT VALUES USING THE SOLVER

Your calculator solves for the input values of any equation that is in the form *"expression = constant"*. This means that all terms involving the variable must be on one side of the equation and constant terms must be on the other side before you enter the equation into the calculator. The expression can, but does not have to, use *x* as the input variable.

The TI-86 offers several methods of solving for input variables. We first illustrate using the SOLVER. (Solving using graphical methods will be discussed after using the SOLVER is explored.) You can refer to an equation that you have already entered in the y(x)= list or you can enter the equation in the solver.

Return to the home screen with [2nd] [EXIT] (QUIT). Access the solver by pressing [2nd] [GRAPH] (SOLVER). If there are no equations stored in the solver, you will see the screen displayed on the right – or	eqn:█
– if the solver has been previously used, you will see a screen similar to the one shown on the right. If this is the case, press ▲ until only the eqn: line is on the screen and [CLEAR] to delete the old equation. You should then have the screen that is shown in the previous step.	exp=2x²−3 exp=0 x=2 bound=(-1E99,1E99)

Let's now use the solver to answer the question in part *e* of Example 4 in Section 1.1: "When did the land value reach \$20,000?" Because the land value is given by $v(t) = 3.622(1.093^t)$ thousand dollars where *t* is the number of years after the end of 1980, we are asked to *solve* the equation $3.622(1.093^t) = 20$. That is, we are asked to find the input value *t* that makes this equation a true statement.

If you already have y1 = 3.622(1.093^ x) in the graphing list, you can refer to the function as y1 in the SOLVER. (Note that y1 can be entered by pressing the F-key under its location in the menu at the bottom of the screen.) If not, enter 3.622(1.093^ x) instead of y1 in the eqn: location. Press [ENTER]. Enter 20 in the exp: location under y1 to tell the TI-86 the rest of the equation.	exp=y1 exp=20 x=2 bound=(-1E99,1E99)
If you need to edit the equation, press ▲ until the previous screen reappears. Edit the equation and then return here. With the cursor in the x location, enter a guess – say 19. (You could have also used as a guess* the value that was in the x location when you accessed the SOLVER.)	exp=y1 exp=20 x=19█ bound=(-1E99,1E99)

*More information on entering a guess appears at the end of this discussion.

CAUTION: You should not change anything in the "bound" location of the SOLVER. The values in that location are the ones between which the TI-86 searches[1] for a solution. If you should accidentally change anything in this location, exit the solver, and begin the entire process again. (The bound is automatically reset when you exit the SOLVER.)

> Be certain that the cursor is on the line corresponding to the input variable for which you are solving (in this example, x). Solve for the input by pressing F5 [SOLVE].
>
> The answer to the original question is that the land value was $20,000 about 19.2 years after 1980 – *i.e.*, in the year 2000.

- Notice the black dot that appears next to x and next to the last line on the above screen. This is the TI-86's way of telling you that a solution has been found. When the bottom line on the screen that states left – rt ≈ 0, the value found for x is an exact solution.

- If a solution continues beyond the edge of the calculator screen, you see "..." to the right of the value. Be certain that you press and hold ▶ to scroll to the end of the number.

  The value may be given in scientific notation, and the portion that you cannot see determines the location of the decimal point. (See *Basic Operation*, #3, in this *Guide*.)

1.1.8 HOW TO DETERMINE A GUESS TO USE IN THE EQUATION SOLVER What you use in the solver as a guess tells the TI-86 where to start looking for the answer. How close your guess is to the actual answer is not very important unless there is more than one solution to the equation. *If the equation has more than one answer, the solver will return the solution that is closest to the guess you supply.* In such cases, you need to know how many answers you should search for and their approximate locations.

Three of the methods that you can use to estimate the value of a guess for an answer from the SOLVER follow. We illustrate these methods using the land value function from Example 4 of Section 1.1 and the equation $v(t) = 3.622(1.093^t) = 20$.

> 1. Enter the function in some location of the graphing list – say y1 = 3.622(1.093^ x) and draw a graph of the function. Press F4 [TRACE] and hold down either ▶ or ◀ until you have an estimate of where the output is 20. Use this *x*-value, 19 or 19.3 or 19.33, as your guess in the SOLVER.

Remember that you can use any letter to represent the input in the solver, but the TI-86 will only graph a function when you use x as the input variable.

> 2. Enter the left- and right-hand sides of the equation in two different locations of the y(x)= list – say y1 = 3.622(1.093^ x) and y2 = 20. With the graph on the screen, press F4 [TRACE] and hold down either ▶ or ◀ until you get an estimate of the X-value where the curve crosses the horizontal line representing 20.

---

[1] It is possible to change the bound if the calculator has trouble finding a solution to a particular equation. This, however, should happen rarely. Refer to the *TI-86 Graphing Calculator Guidebook* for details.

3. Use the AUTO setting in the TABLE, and with ▲ or ▼ scroll through the table until a value near the estimated output is found. Use this *x*-value or a number near it as your guess in the SOLVER. (Refer to Section 1.1.6 of this *Guide* to review the instructions for using the TABLE.)

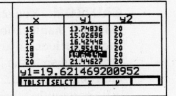

The TI-86 lets you draw a graph from which to estimate a guess for the SOLVER from within the SOLVER. When you use this feature, you are using a combination of all three of the above methods because the calculator draws a graph of *left side of equation − right side of equation* at each *x*-value between xMin and xMax and displays the value of the difference of the two sides of the equation as you trace the graph.

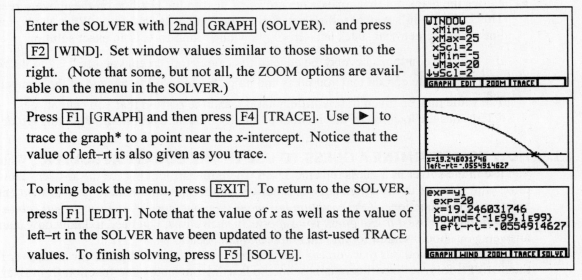

Enter the SOLVER with 2nd GRAPH (SOLVER). and press F2 [WIND]. Set window values similar to those shown to the right. (Note that some, but not all, the ZOOM options are available on the menu in the SOLVER.)

Press F1 [GRAPH] and then press F4 [TRACE]. Use ▶ to trace the graph* to a point near the *x*-intercept. Notice that the value of left−rt is also given as you trace.

To bring back the menu, press EXIT. To return to the SOLVER, press F1 [EDIT]. Note that the value of *x* as well as the value of left−rt in the SOLVER have been updated to the last-used TRACE values. To finish solving, press F5 [SOLVE].

*It is not easy to set a good window for this graph. However, you can still trace the graph, even if you do not see any of it on the screen. If this is the case, carefully check the left − right values to see in which direction to trace. Remember that you are looking for an input where left − right is approximately zero.

1.1.9 GRAPHICALLY FINDING INTERCEPTS

Finding the input value at which the graph of a function crosses the vertical and/or horizontal axis can be found graphically or by using the SOLVER. Remember the process by which we find intercepts:

- To find the *y*-intercept of a function $y = f(x)$, set $x = 0$ and solve the resulting equation.
- To find the *x*-intercept of a function $y = f(x)$, set $y = 0$ and solve the resulting equation.

An *intercept* is the where the graph crosses or touches an axis. Also remember that the *x*-intercept of the function $y = f(x)$ has the same value as the root or solution of the equation $f(x) = 0$. **Thus, finding the *x*-intercept of the graph of $f(x) - c = 0$ is the same as solving the equation $f(x) = c$.**

We illustrate this method with a problem similar to the one in Activity 40 in Section 1.1 of *Calculus Concepts*. You should practice by solving the equation $3.622(1.093^x) = 20$ using this graphical method and by solving the equation that follows using the SOLVER.

Suppose we are asked to find the input value of $f(x) = 3x - 0.8x^2 + 4$ that corresponds to the output $f(x) = 2.3$. That is, we are asked to find *x* such that $3x - 0.8x^2 + 4 = 2.3$. Because this function is not given in a context, we have no indication of an interval of input values to use when drawing the graph. We will use the ZOOM features to set an initial view and then manually set the WINDOW until we see a graph that shows the important points of the function (in this case, the intercept or intercepts.) You can solve this equation graphically using either the *x-intercept method* or the *intersection method*. We present both, and you should use the one you prefer.

X-INTERCEPT METHOD for solving the equation $f(x) - c = 0$:

Press [GRAPH] [F1] [y(x)=] and clear all locations with [CLEAR]. Enter the function $3x - 0.8x^2 + 4 - 2.3$ in y1. You can enter x^2 with [x-VAR] [x^2] or enter it with [x-VAR] [^] 2. Remember to use [−], not [(−)], for the subtraction signs.

NOTE: Whenever there are two menus at the bottom of the display screen, press [EXIT] to delete the bottom menu or press [2nd] before pressing the F-key you want to access a certain command in the top menu. We give instructions assuming there is only one menu.

Draw the graph with [F3] [ZOOM] [MORE] [F4] [ZDECM] or [F4] [ZSTD]. If you use the former, press [F2] [WIND] and reset yMax to 5 to get a better view of the graph. (If you reset the window, press [F5] [GRAPH] to draw the graph.)	
To graphically find an *x*-intercept, *i.e.*, a value of *x* at which the graph crosses the horizontal axis, press [MORE] [F1] [MATH] [F1] [ROOT]. Press and hold [◄] until you are near, but to the *left* of, the leftmost *x*-intercept. Press [ENTER] to mark the location of the *left* bound for the *x*-intercept.	
Notice the small arrowhead (▶) that appears above the location to mark the left bound. Now press and hold [▶] until you are to the *right* of this *x*-intercept. Press [ENTER] to mark the location of the *right* bound for the *x*-intercept.	
For your "guess", press [◄] to move the cursor near to where the graph crosses the horizontal axis. Press [ENTER].	
The input of the leftmost *x*-intercept is displayed as x = −0.5. Note that if this process does not return the correct value for the intercept you are trying to find, you have probably not included the place where the graph crosses the axis between the two bounds (*i.e.*, between the ▶ and ◄ marks on the graph.)	
Repeat the above procedure to find the other *x*-intercept. Confirm that it occurs where x = 4.25.	

INTERSECTION METHOD for solving the equation $f(x) = c$:

Press [GRAPH] [F1] [y(x)=] and clear all locations with [CLEAR]. Enter one side of the equation, $3x - 0.8x^2 + 4$ in y1 and the other side of the equation 2.3, in y2.	
Draw the graphs with [F3] [ZOOM] [MORE] [F4] [ZDECM] or [F4] [ZSTD]. If you use the former, press [F2] [WIND] and reset yMax to 8 to get a better view of the graph. (If you reset the window, press [F5] [GRAPH] to draw the graph.)	
To graphically find where y1 = y2, press [MORE] [F1] [MATH] [MORE] [F3] [ISECT]. Note that the number corresponding to the function's location in the y(x)= list is shown at the top right of the screen. Press [ENTER] to mark the first curve.	
The cursor jumps to the other function – here, the line. Note that the number corresponding to the function's location in the y(x)= list is shown at the top right of the screen. Next, press [ENTER] to mark the second curve.	
Next, supply a guess for the point of intersection. Use [◄] to move the cursor near the intersection point you want to find – in this case, the leftmost point. Press [ENTER].	
The value of the leftmost x-intercept has the x-coordinate $x = -0.5$. Repeat the above procedure to find the rightmost x-intercept. Confirm that it is where $x = 4.25$.	

1.1.10 SUMMARY OF ESTIMATING AND SOLVING METHODS Use the method you prefer.

When you are asked to *estimate* or *approximate* an output or an input value, you can:
- Trace a graph entered in the y(x)= list   (Section 1.1.4)
- Trace a graph showing values of the left – right sides of an equation
  that is entered in the SOLVER   (Section 1.1.8)
- Use close values obtained from the TABLE   (Sections 1.1.3, 1.1.5, 1.1.7)

When you are asked to *find* or *determine* an output or an input value, you should:
- Evaluate an output on the graphics screen   (Section 1.1.3)
- Evaluate an output on the home screen   (Section 1.1.5)
- Evaluate an output value using the table   (Section 1.1.6)
- Find an input using the solver   (Sections 1.1.7, 1.1.8)
- Find an input value from the graphic screen (using the x-intercept
  method or the intersection method)   (Section 1.1.9)

TI-86 Guide

B-13

1.3 Constructed Functions

Your calculator can find output values of and graph combinations of functions in the same way that you do these things for a single function. The only additional information you need is how to enter constructed functions in the graphing list.

1.3.1 FINDING THE SUM, DIFFERENCE, PRODUCT, QUOTIENT OR COMPOSITE FUNCTION
Suppose that a function f has been entered in y1 and a function g has been entered in y2. Your calculator will evaluate and graph these constructed functions:

Enter y1 + y2 in y3 to obtain the sum function $(f + g)(x) = f(x) + g(x)$.

Enter y1 – y2 in y4 to obtain the difference function $(f - g)(x) = f(x) - g(x)$.

Enter y1*y2 in y5 to obtain the product function $(f \cdot g)(x) = f(x) \cdot g(x)$.

Enter y1/y2 in y6 to obtain the quotient function $(f \div g)(x) = \dfrac{f(x)}{g(x)}$.

Enter y1(y2) in y7 to obtain the composite function $(f \circ g)(x) = f(g(x))$.

The functions can be entered in any location in the y(x)= list. Although the TI-86 will not give an algebraic formula for a constructed function, you can check your final answer by evaluating your constructed function and the calculator-constructed function at several different points to see if they yield the same outputs.

1.3.2 FINDING A DIFFERENCE FUNCTION
We illustrate this technique with the functions that are given on page 31 of Section 1.3 of *Calculus Concepts*: Sales = $S(t) = 3.570(1.105^t)$ million dollars t years after 1996 and costs = $C(t) = -39.2t^2 + 540.1t + 1061.0$ thousand dollars t years after 1996.

Press GRAPH F1 [y(x)=], and clear each previously-entered equation with CLEAR or F4 [DELf]. Enter S in y1 with 3 . 570 (1 . 105 ^ x-VAR) ENTER . Enter C in y2 with (-) 39 . 2 x-VAR x^2 + 540 . 1 x-VAR + 1061 ENTER .	Plot1 Plot2 Plot3 \y1■3.570(1.105^x) \y2■-39.2x²+540.1x+1... \y3=■
The difference function, the profit $P(x) = S(x) - 0.001C(x) =$ y1 – 0.001 y2, is entered in y3 with 2nd ALPHA (alpha) 0 (y) 1 – . 001 2nd ALPHA (alpha) 0 (y) 2.	Plot1 Plot2 Plot3 \y1■3.570(1.105^x) \y2■-39.2x²+540.1x+1... \y3■y1-.001y2

- Note that you can also type y when entering the third function in the graphing list by pressing the F-key corresponding to its position on the menu bar.

To find the profit in 1998, evaluate y3 when x = 2. You can evaluate on the home screen, the graphics screen, or in the table. We choose to use the home screen.

Return to the home screen with 2nd EXIT (QUIT). Press 2nd ALPHA (alpha) 0 (y) 3 (2) ENTER to see the result. We find that the profit in 1998 was $P(2) \approx \$2.375$ million.	y3(2) 2.37465925

1.3.3 FINDING A PRODUCT FUNCTION
We illustrate this technique with the functions that are given on page 32 of Section 1.3 of *Calculus Concepts*: Milk price = $S(x) = 0.007x + 1.492$

Copyright © Houghton Mifflin Company. All rights reserved.

dollars per gallon on the xth day of last month and milk sales = $G(x) = 31 - 6.332(0.921^x)$ gallons of milk sold on the xth day of last month.

Press GRAPH F1 [y(x)=], and clear each previously entered equation with CLEAR or F4 [DELf]. Enter S in y1 using . 007 x-VAR + 1 . 492 ENTER and enter G in y2 with 31 − 6 . 332 ((. 921 ^ x-VAR)) ENTER.	Plot1 Plot2 Plot3 \y1■.007 x+1.492 \y2■31-6.332(.921^x) \y3=■
Enter the product function $T(x) = S(x) \cdot G(x)$ = y1·y2 in y3 with the keystrokes 2nd ALPHA (alpha) 0 (y) 1 × 2nd ALPHA (alpha) 0 (y) 2.	Plot1 Plot2 Plot3 \y1■.007 x+1.492 \y2■31-6.332(.921^x) \y3■y1*y2

NOTE: You do not have to use × between y1 and y2 to indicate a product function. You cannot use parentheses to indicate the product function because the TI-86 will think that you are entering y3 as a composite function.

To find milk sales on the 5th day of last month, evaluate y3 at $x = 5$. We choose to do this on the home screen. Return to the home screen with 2nd MODE (QUIT). Press 2nd ALPHA (alpha) 0 (y) 3 ((5)) ENTER to see the result. We find that milk sales were $T(5) \approx \$40.93$.	y3(5) 40.9296552175

1.3.4 CHECKING YOUR ANSWER FOR A COMPOSITE FUNCTION

We illustrate this technique with the functions that are given on page 33 of Section 1.3 of *Calculus Concepts*: altitude = $F(t) = -222.22t^3 + 1755.95t^2 + 1680.56t + 4416.67$ feet above sea level where t is the time into flight in minutes and air temperature = $A(F) = 277.897(0.99984^F) - 66$ degrees Fahrenheit where F is the number of feet above sea level. Remember that when you enter functions in the y(x)= list, you must use x as the input variable.

Clear the functions from the y(x)= list. Enter F in y1 by pressing (−) 222 . 22 x-VAR ^ 3 + 1755 . 95 x-VAR x^2 + 1680 . 56 x-VAR + 4416 . 67 ENTER. Enter A in y2 by pressing 277 . 897 ((. 99984 ^ x-VAR)) − 66 ENTER.	Plot1 Plot2 Plot3 \y1■-222.22x^3+1755... \y2■277.897(.99984^x... \y3=■
Enter the composite function $(A \circ F)(x) = A(F(x))$ = y2(y1) in y3 with 2nd ALPHA (alpha) 0 (y) 2 ((2nd ALPHA (alpha) 0 (y) 1)) ENTER.	Plot1 Plot2 Plot3 \y1■-222.22x^3+1755... \y2■277.897(.99984^x... \y3■y2(y1) \y4=■
Next, enter your algebraic answer for the composite function in y4. Be certain that you enclose the exponent in y4 (the function in y1) in parentheses! (The composite function in the text is the one that appears to the right, but you should enter the function that you have found for the composite function.)	Plot1 Plot2 Plot3 \y1■-222.22x^3+1755... \y2■277.897(.99984^x... \y3■y2(y1) \y4■277.897(.99984^(...

We now wish to check that the algebraic answer for the composite function is the same as the calculator's composite function by evaluating both functions at several different input values. You can do these evaluations on the home screen, but using the table is more convenient.

1.3.5 TURNING FUNCTIONS IN THE GRAPHING LIST OFF AND ON Note that we are interested in the output values for only y3 and y4. However, the table will list values for all functions that are turned on. (A function is *turned on* when the equals sign in its location in the graphing list is darkened.) We now wish to turn off y1 and y2.

> Press GRAPH F1 [y(x)=] and place the cursor in the line containing y1. Press F5 [SELCT] and use ▼ to move the cursor to the y2 line. Press F5 [SELCT]; y1 and y2 are now turned off.

A function is *turned off* when the equals sign in its location in the graphing list is not dark. To turn a function back on, simply reverse the above process to make the equal sign for the function dark. The SELCT key toggles between the function being off and being on. When you draw a graph, the TI-86 graphs of all functions that are turned on. You may at times wish to keep certain functions entered in the graphing list but not have them graph and not have their values shown in the table. Such is the case in this illustration. We now return to checking to see that y3 and y4 represent the same function.

> Choose the ASK setting in the table setup so that you can check several different values for both y3 and y4. Recall that you access the table setup with TABLE F2 [TBLST]. Move the cursor to ASK in the Indpnt: location and press ENTER.

> Press F1 (TABLE), type in the *x*-value(s) at which the function is to be evaluated, and press ENTER after each one. We see that because all these outputs are the same for each function, you can be fairly sure that your answer is correct.

Why does ERROR appear in the table when $x = 57$? Look at the value when $x = 20$; it is very large! The computational limits of the calculator have been exceeded when $x = 57$. The TI-86 calls this an OVERFLOW error.

1.3.6 GRAPHING AN INVERSE FUNCTION The TI-86 can draw the graph of the inverse of a function. Using the calculator graph, you can check your algebraic answer. We illustrate this idea using the function in Example 4 of Section 1.3 of *Calculus Concepts*:

> The fares for a cab company are determined by the function $F(d) = 1.8d + 2.5$ dollars where d is the distance traveled in miles.

> Press GRAPH F1 [y(x)=], clear any previously entered functions, and enter the function F in y1. Remember to use x as the input variable. (Refer to Sections 1.1.2 and 1.1.3 of this *Guide* for hints on how to set the window to graph this function.)

This problem does not state an interval of values for which the input variable is defined, so we begin by guessing one that makes sense in context. Because d is the distance traveled, we know that $d \geq 0$. We choose a maximum value of 50. (Your guess is as good as the one that is given below.)

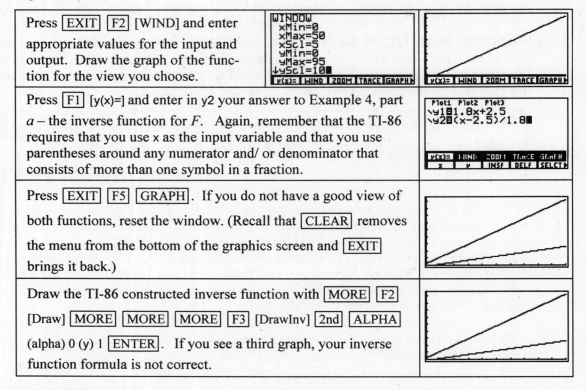

Press EXIT F2 [WIND] and enter appropriate values for the input and output. Draw the graph of the function for the view you choose.	
Press F1 [y(x)=] and enter in y2 your answer to Example 4, part a – the inverse function for F. Again, remember that the TI-86 requires that you use x as the input variable and that you use parentheses around any numerator and/or denominator that consists of more than one symbol in a fraction.	
Press EXIT F5 GRAPH. If you do not have a good view of both functions, reset the window. (Recall that CLEAR removes the menu from the bottom of the graphics screen and EXIT brings it back.)	
Draw the TI-86 constructed inverse function with MORE F2 [Draw] MORE MORE MORE F3 [DrawInv] 2nd ALPHA (alpha) 0 (y) 1 ENTER. If you see a third graph, your inverse function formula is not correct.	

CAUTION: If you do not see a third graph, as is the case in the graph shown above, you may have computed the formula correctly. However, you might have an incorrect inverse function formula whose graph does not show in the current viewing window. We suggest that if you do not see a third graph that you turn off the graphing location where you have entered your inverse function formula – here y2. (Remember that you turn off a function by having the cursor on that function location and pressing F5 [SELCT].) Then, on the home screen, press 2nd ENTER (ENTRY) ENTER to redraw the inverse function graph, and visually check that it is the same graph as the one for y2. Realize that these methods provide only a visual check on your answer and are not exact.

The answers to parts b and c of Example 4 are found using the distance (inverse) function in y2 and the fare (original) function in y1, respectively. Don't forget to include units of measure with the answers.	

1.3.7 COMPOSITION PROPERTY OF INVERSE FUNCTIONS
This concept, involved in part d of Example 4, provides another check on your answer for the inverse function.

Form the compositions of the inverse function and the original function according to the Composition Property of Inverse Functions and enter them in y3 and y4. Turn off the functions in y1 and y2.	

Have the TABLE set to ASK mode (See Section 1.1.6) and press TABLE F1 [TABLE]. Enter several values for x to see if both composite functions return that value of x. If so, your answer for the inverse function is very likely correct.

1.3.8 GRAPHING A PIECEWISE CONTINUOUS FUNCTION

Piecewise continuous functions are used throughout the text. You will need to use your calculator to graph and evaluate outputs of piecewise continuous functions. Several methods can be used to draw the graph of a piecewise function. One of these is presented below using the function that appears in Example 6 of Section 1.3 in *Calculus Concepts*:

The population of West Virginia from 1985 through 1993 can be modeled by

$$P(t) = \begin{cases} -23.514t + 3903.667 \text{ thousand people} & \text{when } 85 \le t < 90 \\ 7.7t + 1098.7 \text{ thousand people} & \text{when } 90 \le t \le 93 \end{cases}$$

where t is the number of years since 1900.

Parentheses are used around each portion of the function and around an inequality statement that tells the calculator which side of the break point to graph each part of the function. We set the lower and upper endpoints of the input interval as xMin and xMax, respectively. Clear any functions that are in the y(x)= list.

Enter the first piece of P (using x as the input variable) in the y1 location with the keystrokes [(] [(-)] 23.514 x-VAR [+] 3903.667 [)] [(] x-VAR 2nd 2 (TEST) F2 [<] 90 [)] ENTER.

Enter the second piece of the function P (using x as the input variable) in the y2 location with [(] 7.7 x-VAR [+] 1098.7 [)] [(] x-VAR F5 [≥] 90 [)].

Each piece of the function and its corresponding input at the break point must be inside parentheses.

Your calculator draws graphs by connecting function outputs wherever the function is defined. However, this piecewise function breaks at $x = 90$. The TI-86 will connect the two pieces of P unless you tell it not to do. Whenever you draw graphs of piecewise functions, you should set your calculator to Dot mode as described below so that it will not connect the different pieces of the function.

Press GRAPH F1 [y(x)=] and then press MORE F3 [STYLE]. Press F3 five more times and the slanted line[2] to the *left* of y1 should be a dotted line (as shown to the right). Press ▼ and then press F3 six times to put the function in y2 in Dot mode.

Note: The method described above places individual functions in Dot mode. The functions return to standard (connected) mode when the function locations are cleared. If you want to put all functions in Dot mode at the same time, press GRAPH F2 [WIND] MORE F3 [FORMT], choose DrawDot in the third line, and press ENTER. However, if you choose to

---

[2] The different graph styles you can draw from this location are described in more detail on page 79 in your *TI-86 Graphing Calculator Guidebook*.

set DOT mode in this manner, you must return to the window format screen, select DrawLine instead of DrawDot, and press ENTER to take the TI-86 out of DOT mode.

| Now, set the window. The function P is defined only when the input is between 85 and 93. So, we evaluate $P(85)$, $P(93)$, and $P(90)$ to help when setting the vertical view. | y1(85) 1904.977
y2(90)
y2(93) 1791.7
        1814.8 |

- If you attempt to set the window using ZFIT as described in Section 1.1.2, the picture is not very good and you need to manually reset the height of the window as described below.

Press GRAPH F2 [WIND], set xMin = 85, xMax = 93, yMin ≈ 1780, and yMax ≈ 1905. Press F5 [GRAPH] CLEAR.	
Reset the window if you want a closer look at the function around the break point. The graph to the right was drawn using xMin = 89, xMax = 91, yMin = 1780, and yMax = 1810.	
You can find function values by evaluating outputs on the home screen or using the table. Carefully look at the inequalities in the function P to determine whether y1 or y2 should be evaluated to obtain each particular output. (See Sections 1.1.5 and 1.1.6 for more detailed instructions evaluating functions.)	y1(87) 1857.949 y2(92) y1(85) 1807.1 y2(90)■ 1904.977

1.4 Limits: Describing Function Behavior

The TI-86 table is an essential tool when you are estimating limits numerically. Even though rounded values are shown in the table due to space limitations, the TI-86 displays at the bottom of the screen many more decimal places for a particular output when you highlight that output.

1.4.1 ESTIMATING LIMITS NUMERICALLY Whenever you use the TI-86 to estimate limits numerically, you will find it easiest to set the TABLE to ASK mode. We illustrate using the function $u(x) = \dfrac{3x^2 + 3x}{9x^2 + 11x + 2}$ that appears in Example 2 of Section 1.4 in *Calculus Concepts*:

Press GRAPH F1 [y(x)=] and use F4 [DELf] to delete all previously-entered functions. Enter u, being certain to enclose both numerator and denominator of the fraction in parentheses.	
Press TABLE F2 [TBLST]. Choose Ask in the Indpnt: location by placing the cursor over Ask and pressing ENTER.	
Press F1 [TABLE]. Delete any values that are left from prior use with DEL. (*Note*: The values you enter do not have to be those shown in the text or these shown to the right.)	

We are numerically estimating $\lim_{x \to {}^-1^-} u(x)$. This is why, on the last table screen, we entered values to the *left* of, and becoming *closer and closer* to, -1.

CAUTION: Your instructor will very likely have you write the table you construct on paper. The TI-86 eventually rounds off the table values it displays because of space limitations. Be certain to highlight the values and look on the bottom of the screen to see what these values actually are.

ROUNDING OFF: Recall that *rounded off* means that if one digit past the digit of interest if less than 5, other digits past the digit of interest are dropped. If one digit past the one of interest is 5 or over, the digit of interest is increased by 1 and the remaining digits are dropped. (For instance, the highlighted number in the last screen shown above, -1.000005, has been rounded off by the TI-86 to 5 decimal places. Because the 5 in the sixth decimal place location is in the "5 or over" category, the 0 before this 5 is increased by 1 with the digits past that point dropped to give -1.00001.)

RULE OF THUMB FOR DETERMINING LIMITS FROM TABLES: Suppose that we want $\lim_{x \to {}^-1^-} u(x)$ accurate to 3 decimal places. Watch the table until you see that the output is the same value to one more decimal place (here, to 4 decimal places) for three consecutive outputs. Then, round that common value off to the requested 3 places for the desired limit. Your instructor may establish a different rule from this one, so be sure to ask.

Using this Rule of Thumb and the results that are shown on the last calculator screen, we estimate that $\lim_{x \to {}^-1^-} u(x) = 0.429$. We now need to estimate the limit from the right of -1. (*Note*: Again, the values that you enter do not have to be those shown in the text or the ones shown below.)

Delete the values currently in the table with $\boxed{\text{DEL}}$. To numerically estimate $\lim_{x \to {}^-1^+} u(x)$, enter values to the *right* of, and becoming *closer and closer* to, -1. Because the output $0.4285\ldots$ appears three times in a row, we estimate $\lim_{x \to {}^-1^+} u(x) = 0.429$.

Because $\lim_{x \to {}^-1^-} u(x) = \lim_{x \to {}^-1^+} u(x) = 0.429$, we estimate that $\lim_{x \to {}^-1} u(x) = 0.429$. We now illustrate finding the limit in part *b* of Example 2 in Section 1.4 of *Calculus Concepts*:

Delete the values currently in the table with $\boxed{\text{DEL}}$. To numerically estimate $\lim_{x \to {}^-2/9^-} u(x)$, enter values to the *left* of, and becoming *closer and closer* to, $-2/9 = -0.222222\ldots$.

Because the outputs appear to be getting larger and larger, we estimate $\lim_{x \to {}^-2/9^-} u(x) \to \infty$.

Delete the values currently in the table. To numerically estimate $\lim_{x \to {}^-2/9^+} u(x)$, enter values to the *right* of, and getting *closer and closer* to, $-2/9$.

Because the outputs appear to become larger and larger, we estimate $\lim_{x \to {}^-2/9^+} u(x) \to -\infty$. Thus, $\lim_{x \to {}^-2/9} u(x)$ does not exist.

1.4.2 CONFIRMING LIMITS GRAPHICALLY – ZOOMING IN AND OUT

A graph can be used to confirm a limit that you estimated numerically. You also can *zoom in* or *zoom out* on the graph to obtain a better view of the limit you are estimating. We again illustrate using the function $u(x) = \dfrac{3x^2 + 3x}{9x^2 + 11x + 2}$ that is in Example 2 of Section 1.4 in *Calculus Concepts*.

Have the function u entered in the y1 location of the y(x)= list. A graph drawn with [GRAPH] [F3] [ZOOM] [F4] [ZSTD] or with [GRAPH] [F3] [ZOOM] [MORE] [F4] [ZDECM] is not very helpful. (Refer to Sections 1.1.2 and 1.1.3 of this *Guide* for hints on how to set the window to graph this function.)	
To confirm that $\lim\limits_{x \to -1} u(x) = 0.429$, we are only interested in values of u that are near -1. So, choose values very near to -1 for the *x*-view and evaluate the function at those *x*-values to help determine the *y*-view. We manually set the window to values such as those shown to the right and draw the graph.	
If you look closely, you can actually see the "hole" in the graph at $x = -1$. Press [F4] [TRACE] in this window, the TI-86 confirms that u is not defined at $x = -1$.	
After pressing [F4] [TRACE], press [▶] and observe the *y*-value. Press [◀] several times to go to the left of $x = -1$ and confirm from the *y*-values that the limit is approximately what we had determined numerically.	

Note that we confirmed that the limit exists by seeing that the two parts of the graph (to the left and right of $x = -1$) move toward the same point. Tracing around $x = -1$ provides a check on the numerical value of the limit.

The previous instructions show how to *zoom in* by manually setting the window. You can also zoom in with the ZOOM menu of the calculator. We next describe this method.

Return to the graphing screen set with [GRAPH] [F3] [ZOOM] [MORE] [F4] [ZDECM] (or any screen where you can see the portion of the graph around $x = -1$.) Press [F3] [ZOOM] [F2] [ZIN] and use [◀] and [▲] to move the blinking cursor until you are near the point on the graph where $x = -1$.	
Press [ENTER]. If you look closely, you can see the hole in the graph at $x = -1$. If this view is not magnified enough to see what is happening around $x = -1$, press [ENTER] to zoom in again.	

TI-86 Guide   B-21

| To check your numerical estimate of the limit, press [EXIT] [F4] [TRACE], use the arrow keys to move to either side of $x = -1$, and observe the y-values. | |

We now consider the limit in part *b* of Example 2 of Section 1.4. We want to confirm with a graph what was found numerically; that is, we wish to confirm that $\lim_{x \to -2/9} u(x)$ does not exist.

However, first we should eliminate the "extra" vertical line that appears on the above graph at $x = -2/9$. This line appears because we are graphing in DrawLine mode (which tells the TI-86 to connect points on the graph). Note that you may or may not have the line on your graph.

Have y1 set to draw in Dot mode. (See page B-17 for the keystrokes needed to put the TI-86 in DrawDot mode.) If you use [F3] in the ZOOM menu (ZOUT), the graph is lost. We therefore manually set the window.	
We want to zoom out on the graph to confirm that the limit of $u(x)$ as x approaches $-2/9$ does not exist. To do this, we set a very small x-view and a larger y-view. (Note that these values are arbitrary – any small x-view that includes $-2/9$ and any y-view in which the graph can be seen clearly will do.)	
You can TRACE the graph to confirm that where it "splits" is near $x = -2/9 \approx -0.222222...$. It appears that as x approaches $-2/9$ from the left that the outputs become larger and larger and that as x approaches $-2/9$ from the right that the outputs become smaller and smaller.	

1.4.3 INVESTIGATING END BEHAVIOR NUMERICALLY

Investigating end behavior using the TI-86 table is very similar to numerically estimating the limit at a point. We illustrate this using the function that appears in Example 4 of Section 1.4 in *Calculus Concepts*.

Have $u(x) = \dfrac{3x^2 + x}{3x + 11x^2 + 2}$ in the y1 location of the y(x)= list. Be certain that you remember to enclose both the numerator and denominator of the fraction in parentheses.	
Have TBLST set to Ask and press [TABLE] [F1] [TABLE]. In order to numerically estimate $\lim_{x \to \infty} u(x)$, enter values of x that get larger and larger. (*Note*: The values that you enter do not have to be those shown in the text or these shown to the right.)	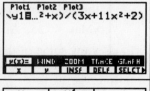
We assume that we want the limit accurate to 3 decimal places. According to our *Rule of Thumb for Determining Limits from Tables* (page B-19), once we see the same value to 4 decimal positions three times in a row, we can estimate the limit by rounding off the answer to 3 decimal places.	$\lim_{x \to \infty} u(x) \approx 0.273$ Note that the last entry in the table ($x = 10{,}000$) was not necessary.

Part *b* of Example 4 asks for $\lim_{x \to -\infty} u(x)$. Delete the values that are currently in the table with [DEL]. Then, enter values of *x* that get smaller and smaller. We estimate $\lim_{x \to -\infty} u(x) \approx 0.273$.

CAUTION: It is not the final value, but a sequence of several values, that is important when determining limits. If you enter a very large or very small value, you may exceed the limits of the TI-86's capability and obtain an incorrect number. Always look at the sequence of values obtained to make sure that all values found make sense.

1.4.4 INVESTIGATING END BEHAVIOR GRAPHICALLY As was the case with limits at a point, a graph of the function can be used to confirm a numerically-estimated limit. We again illustrate with the function that appears in Example 4 of Section 1.4 in *Calculus Concepts*.

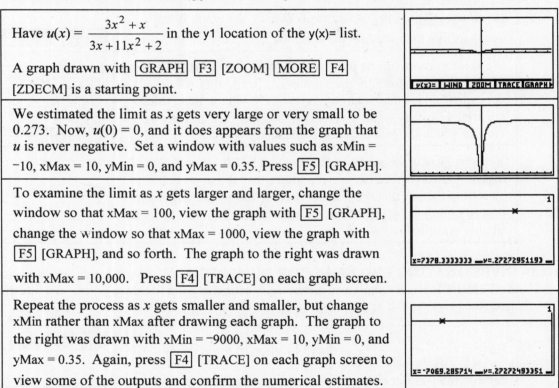

Have $u(x) = \dfrac{3x^2 + x}{3x + 11x^2 + 2}$ in the y1 location of the y(x)= list. A graph drawn with [GRAPH] [F3] [ZOOM] [MORE] [F4] [ZDECM] is a starting point.	
We estimated the limit as *x* gets very large or very small to be 0.273. Now, $u(0) = 0$, and it does appears from the graph that *u* is never negative. Set a window with values such as xMin = ⁻10, xMax = 10, yMin = 0, and yMax = 0.35. Press [F5] [GRAPH].	
To examine the limit as *x* gets larger and larger, change the window so that xMax = 100, view the graph with [F5] [GRAPH], change the window so that xMax = 1000, view the graph with [F5] [GRAPH], and so forth. The graph to the right was drawn with xMax = 10,000. Press [F4] [TRACE] on each graph screen.	
Repeat the process as *x* gets smaller and smaller, but change xMin rather than xMax after drawing each graph. The graph to the right was drawn with xMin = ⁻9000, xMax = 10, yMin = 0, and yMax = 0.35. Again, press [F4] [TRACE] on each graph screen to view some of the outputs and confirm the numerical estimates.	

1.5 Linear Functions and Models

This portion of the *Guide* gives instructions for entering real-world data into the calculator and finding familiar function curves to fit that data. You will use the beginning material in this section throughout all the chapters in *Calculus Concepts*.

CAUTION: Be very careful when you enter data in your calculator because your model and all of your results depend on the values that you enter! Always check your entries.

1.5.1 ENTERING DATA We illustrate data entry using the values in Table 1.46 of Section 1.5.

Year	1996	1997	1998	1999	2000	2001
Tax (in dollars)	2532	3073	3614	4155	4696	5237

TI-86 Guide   B-23

Press [2nd] [+] (STAT) [F2] [EDIT] to access the lists that hold data. You see only the first three lists, (L1, L2, and L3) but you can access the other two lists (L4 and L5) by having the cursor on the list name and pressing [▶] several times. If you do not see L1, L2, and so forth, return to the statistical setup instructions on page B-1 of this *Guide*.

In this text, we usually use list L1 for the input data and list L2 for the output data. If there are any data already in your lists, see Section 1.5.3 of this *Guide* and first delete these "old" data values. To enter data in the lists, do the following:

Position the cursor in the first location in list L1. Enter the input data into list L1 by typing the years from top to bottom in the L1 column, pressing [ENTER] after each entry.

After typing the sixth input value, 2001, use [▶] to cause the cursor go to the top of list L2. Enter the output data into list L2 by typing the entries from top to bottom in the L2 column, pressing [ENTER] after each tax value.

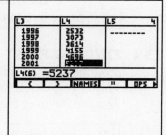

Data can also be entered from the home screen rather than in the statistical lists (as was shown above). The end result is the same, so you should choose the method that you prefer.

On the home screen, press [2nd] [−] (LIST) [F1] [{], and then type in each of the input data separated by commas. End the list with [F2] [}]. Store this list to the name L1 with [STO▶] 7 (L) [ALPHA] 1 [ENTER]. Repeat the process to enter the output data, but store these data in a list named L2.

WARNING: If you do not enter and store the input data into a list named L1 and the output data into a list named L2, many of the programs in this *Guide* will not properly execute.

1.5.2 EDITING DATA No matter how it is entered, the easiest way to edit data is using the statistical lists. If you incorrectly type a data value, use the cursor keys (*i.e.*, the arrow keys [▶], [◀], [▲], and/or [▼]) to darken the value you wish to correct and then type the correct value. Press [ENTER].

- To *insert* a data value, put the cursor over the value that will be directly below the one you will insert, and press [2nd] [DEL] (INS). The values in the list below the insertion point move down one location and a 0 is filled in at the insertion point. Type the data value to be inserted over the 0 and press [ENTER]. The 0 is replaced with the new value.

- To *delete* a single data value, move the cursor over the value you wish to delete, and press [DEL]. The values in the list below the deleted value move up one location.

1.5.3 DELETING OLD DATA Whenever you enter new data in your calculator, you should first delete any previously-entered data. There are several ways to do this, and the most convenient method is illustrated below.

Access the data lists with [2nd] [+] (STAT) [F2] [EDIT]. (You probably have different values in your lists if you are deleting "old" data.) Use [▲] to move the cursor over the name L1.
Press [CLEAR] [ENTER]. Use [▶] and [▲] to move the cursor over the name L2. Press [CLEAR] [ENTER]. Repeat this procedure to clear the other lists.

1.5.4 FINDING FIRST DIFFERENCES When the input values are evenly spaced, you can use program DIFF to compute first differences in the output values. Program DIFF is given in the *TI-86 Program Appendix* at the *Calculus Concepts Website*. Consult the Programs category in *TroubleShooting the TI-86* in this *Guide* if you have questions about obtaining the programs.

Have the data given in Table 1.46 in Section 1.5 of *Calculus Concepts* entered in your calculator. (See Section 1.5.1 of this *Guide*.) Exit the list menu with [2nd] [EXIT] (QUIT).
To run the program, press [PRGM] [NAMES] followed by the F-key that is under the DIFF program location, and press [ENTER]. The message on the right appears on your screen.
If you have not entered the data in L1 and L2, press [F2] [Quit] and do so. Otherwise, press [F1] [Yes] to continue. Press [F1] [1st] to compute the first differences. Choose [F4] to quit.	We use the other choices in the next chapter.	You can also view the first differences in list L3.

- The first differences are constant, so a linear function gives a perfect fit to these tax data.

You will find program DIFF very convenient to use in the next chapter because of other options it has. However, if you do not want to make use program DIFF at this time, you can use a built-in capability of your TI-86 to compute the first differences of any list.

Be certain that you have the output data entered in list L2 in your calculator. Press [2nd] [−] (LIST) [F5] [OPS] [MORE] [MORE] [F4] [Delta1] [ALPHA] 7 (L) 2 [ )] [ENTER].

For larger data sets use [▶] to scroll to the right to see the remainder of the first differences. Use [◀] to scroll back to the left.

Note: Program DIFF **should not** be used for data with input values (L1) that are *not* evenly spaced. First differences give no information about a possible linear fit to data with inputs that are not the same distance apart. If you try to use program DIFF with input data that are not evenly spaced, the message INPUT VALUES NOT EVENLY SPACED appears and the program stops. The Deltalst option discussed above gives first differences even if the input values are not evenly spaced. However, these first differences have no interpretation when finding a linear function to fit data if the input values are not evenly spaced.

1.5.5 SCATTER PLOT SETUP The first time that you draw a graph of data, you need to set the TI-86 to draw the type of graph you want to see. Once you do this, you never need to do this set up again (unless for some reason the settings are changed). If you always put input data in list L1 and output data in list L2, you can turn the scatter plots off and on from the y(x)= screen rather than the STAT PLOTS screen after you perform this initial setup.

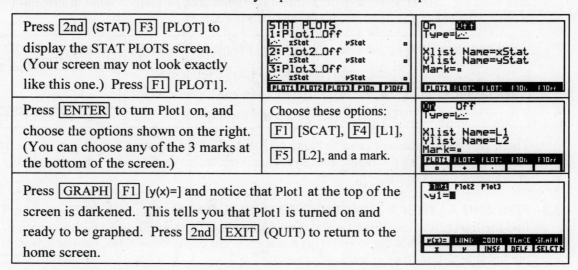

| Press [2nd] (STAT) [F3] [PLOT] to display the STAT PLOTS screen. (Your screen may not look exactly like this one.) Press [F1] [PLOT1]. |
| Press [ENTER] to turn Plot1 on, and choose the options shown on the right. (You can choose any of the 3 marks at the bottom of the screen.) | Choose these options: [F1] [SCAT], [F4] [L1], [F5] [L2], and a mark. |
| Press [GRAPH] [F1] [y(x)=] and notice that Plot1 at the top of the screen is darkened. This tells you that Plot1 is turned on and ready to be graphed. Press [2nd] [EXIT] (QUIT) to return to the home screen. |

- A scatter plot is *turned on* when its name on the y(x)= screen is darkened. From now on, you can turn the scatter plot off and on by having Plot1 either not dark or dark. To turn Plot1 off, use [▲] to move the cursor to the Plot1 location, and press [ENTER]. Reverse the process to turn Plot1 back on.

- You can enter the names of any data lists as the Xlist Name and the Ylist Name and draw a scatter plot. (However, it is easiest to always work with L1 and L2.) All data lists on the TI-86 can be named and stored in the calculator's memory for later recall and use. Refer to Section 1.5.13 and 1.5.14 of this *Guide* for instructions on storing data lists and later recalling them for use.

1.5.6 DRAWING A SCATTER PLOT OF DATA Any functions that are turned on in the y(x)= list will graph when you plot data. Therefore, you should clear, delete, or turn off all functions before drawing a scatter plot. We illustrate how to graph data using the modified tax data that follows Example 2 in Section 1.5 of *Calculus Concepts*.

Year	1996	1997	1998	1999	2000	2001
Tax (in dollars)	2541	3081	3615	4157	4703	5242

Access the y(x)= graphing list. Be sure that Plot 1 at the top left of this screen is darkened.

| Press [2nd] [+] (STAT) [F2] [EDIT]. Using the table given above, enter the year data in L1 and the modified tax data in L2 according to the instructions given in Section 1.5.1 of this *Guide*. (You can either leave values in the other lists or clear them.) | |

Press GRAPH F3 [ZOOM] MORE F5 [ZDATA] to have the TI-86 set an autoscaled view of the data and draw the scatter plot. Note that ZDATA does not reset the *x*- and *y*-axis tick marks. You should do this manually with the window settings if you want different spacing between the tick marks.)

Recall that if the data are perfectly linear (that is, every data point falls on the graph of a line), the first differences in the output values are constant. The first differences for the original tax data were constant at $541, so a linear function fit the data perfectly. What information is given by the first differences for these modified tax data?

Run program DIFF by pressing PRGM F1 [NAMES] followed by F-key under DIFF, and press ENTER. Press F4 [Quit] to exit the program. View the first differences on the home screen or press 2nd + (STAT) F2 [EDIT] to view them in list L3.

These first differences are close to being constant. This information, together with the straight line pattern shown by the scatter plot, are a good indication that a linear function is likely to give a good fit to the data.

1.5.7 FINDING A LINEAR FUNCTION TO MODEL DATA You will often have your TI-86 find the linear function of the form $y = a + bx$ that best fits a set of data. In this equation, a is the *y*-intercept and b is the slope. This equation is not of the same form, but is equivalent to, the linear equation form given in your text.

Press 2nd EXIT (QUIT) to return to the home screen. Then, press 2nd + (STAT) F1 [CALC] F3 [LinR] ALPHA 7 (L) 1 , ALPHA 7 (L) 2 , 2nd ALPHA (alpha) 0 (y) 1.

The keystrokes above find the linear equation of best fit using L1 as the input data, L2 as the output data, and pastes the equation in y1. Press ENTER.

The linear equation of best fit for the modified tax data that was entered into lists L1 and L2 in Section 1.5.6 is displayed on the home screen and has been copied into the y(x)= list. (Remember that the equation scrolls to the right when you press ▶.)

CAUTION: The best-fit function found by the calculator is also called a *regression function*. The coefficients of the regression function ***never*** ***should be rounded***! This is not a problem because the calculator pastes the entire equation it finds into the graphing list at the same time the function is found if you follow the instructions given above.

NOTE: The TI-86 will use lists called xStat and yStat, which probably contain different data, if you do not specify lists L1 and L2 in the instruction to find the best-fit equation. To use other lists to hold the input and output data, just replace L1 and L2 by the other lists in the fit instruction. To paste the function into a location other than y1, just change the number 1 following y in the fit instructions to the number corresponding to the graphing location that you want.

CAUTION: The *r* that is shown on the screen that first gives the linear equation is called the *correlation coefficient*. This and a quantity called r^2, the *coefficient of determination*, are numbers that you will learn about in a statistics course. It is not appropriate[3] to make use of these values in a calculus course.

Graphing the Line of Best Fit: After finding a best-fit equation, you should always draw the graph of the function on a scatter plot to verify that the function provides a good fit to the data.

Press EXIT F5 [GRAPH] to overdraw the function you pasted in the graphing list on the scatter plot of the data.

(As we suspected from looking at the scatter plot and the first differences, this function provides a very good fit to the data.)

1.5.8 USING A MODEL FOR PREDICTIONS
You can use one of the methods described in Sections 1.1.5 or 1.1.6 of this *Guide* to evaluate the linear function at the indicated input value.

Predict the tax owed in 2002 where the tax is found using the linear model computed in the last section of this *Guide:* Tax = $540.371t - 1,076,042.467$ dollars where *t* is the year. Remember that you should always use the full model, *i.e.*, the function you pasted in y1, not the rounded equation given above, for all computations.

The 2002 tax, predicted by evaluating an output on the home screen, is about $5781. We now predict the tax in 2003 using the TI-86 table. As seen to the right, the predicted tax is approximately $6322.

1.5.9 COPYING A GRAPH TO PAPER
Your instructor may ask you to copy what is on your graphics screen to paper. If so, use the following ideas to more accurately perform this task.

After using a ruler to place and label a scale (*i.e.*, tick marks) on your paper, use the trace values (as shown below) or the data to graph a scatter plot on your paper.

Press GRAPH to return the modified tax data graph found in Section 1.5.7 to the screen. Press F4 [TRACE] and ▶. The symbol P1 in the upper right-hand corner of the screen indicates that you are tracing the scatter plot of the data in Plot 1.

Press ▼ to move the trace cursor to the linear function graph. The number in the top right of the screen tells you the location of the function that you are tracing (in this case, y1). Use ▶ and/or ◀ to locate values that are as "nice" as possible and mark those points on your paper. Use a ruler to connect these points and draw the line.

---

[3] Unfortunately, there is no single number that can be used to tell whether one function better fits data than another. The correlation coefficient only compares linear fits and should not be used to compare the fits of different types of functions. For the statistical reasoning behind this statement, read the references in footnote 6 on page A-33.

- If you are copying the graph of a continuous curve rather than a straight line, you need to trace as many points as necessary to see the shape of the curve while marking the points on your paper. Connect the points with a smooth curve.

1.5.10 ALIGNING DATA We return to the modified tax data entered in Section 1.5.6 of this *Guide*. If you want L1 to contain the number of years after a certain year instead of the actual year, you need to *align* the input data. In this illustration, we shift all of the data points to 3 different positions to the left of where the original values are located.

Press [2nd] [+] (STAT) [F2] [EDIT] to access the data lists. To copy the contents of one list to another list; for example, to copy the contents of L1 to L3, use [▲] and [▶] to move the cursor so that L3 is highlighted. Press [ALPHA] 7 (L) 1 [ENTER].

Note: This first step shown above is not necessary, but it will save you the time it takes to re-enter the input data if you make a mistake. Also, it is not necessary to first clear L3. However, if you want to do so, have the symbols L3 highlighted and press [CLEAR] [ENTER].

To align the input data as the number of years past 1996, first press the arrow keys ([◀] and [▲]) so that L1 is highlighted. Tell the TI-86 to subtract 1996 from each number in L1 with [ALPHA] 7 (L) 1 [−] 1996.

Press [ENTER]. Instead of an actual year, the input now represents the number of years since 1996.

Return to the home screen with [2nd] [EXIT] (QUIT).

Find the linear function by pressing [2nd] [ENTER] (ENTRY) as many times as needed until you see the linear fit instruction. To enter this function in a different location, say y2, press [◀] and 2.

Press [ENTER] and then press [GRAPH] [F1] [y(x)=] to see the function pasted in the y2 location.

Note: If you want the aligned function to be in y1, do not replace y1 with y2 before pressing [ENTER] to find the equation.

To graph this equation on a scatter plot of the aligned data, first turn off the function in y1 (see page B-15 of this *Guide*). Press [EXIT] [F3] [ZOOM] [MORE] [F5] [ZDATA].

If you now want to find the linear function that best fits the modified tax data using the input data aligned another way, say as the number of years after 1900, first return to the data lists with [2nd] [+] (STAT) [F2] [EDIT] and highlight L1.

TI-86 Guide   B-29

| Add 96 to each number currently in L1 with ALPHA 7 (L) 1 + 96 ENTER. Instead of an actual year, the input now represents the number of years since 1900. | |

- There are many ways that you can enter the aligned input into L1. Another method that you can use is to start over from the beginning. Replace L1 with the contents of L3 by highlighting L1 and pressing ALPHA 7 (L) 3 ENTER. Once again highlight the symbols L1 and subtract 1900 from each number in L1 with ALPHA 7 (L) 1 − 1900.

| On the home screen, find the linear function for the aligned data by pressing 2nd ENTER (ENTRY) until you see the linear regression instruction. To enter this new equation in a different location, say y3, press ◄ 3 ENTER. Press GRAPH F1 [y(x)=] to see the function pasted in the y3 location. | |

- To graph this equation on a scatter plot of the aligned data, first turn off the other functions and then press EXIT F3 [ZOOM] MORE F5 [ZDATA].

- Remember, if you have aligned the data, the input value at which you evaluate the function may not be the value given in the question you are asked. You can use any of the equations to evaluate function values.

| Using the function in y1 (the input is the year), in y2 (the input is the number of years after 1996), or in y3 (the input is the number of years after 1900), we predict that the tax owed in 2003 is approximately $6322. | y1(2003) 6321.5047619
y2(7)   6321.5047619
y3(103)   6321.5047619 |

1.5.11 NAMING AND STORING DATA You can name data (either input, output, or both) and store it in the calculator memory for later recall. You may or may not want to use this feature; it would be helpful if you plan to use large data sets several times. To illustrate the procedure, let's name the output data.

| Press 2nd EXIT (QUIT) to return to the home screen. You can view any list from the STAT EDIT mode (where the data is originally entered) or from the home screen. View the modified tax data in L2 by pressing ALPHA 7 (L) 2 ENTER. | L2
{2541 3081 3615 4157... |
| Pressing ► allows you to scroll through the list to see the portion that is not displayed. To store this data with the name TAX, press ALPHA 7 (L) 2 STO▶ TAX ENTER. | L2
{2541 3081 3615 4157...
L2→TAX
{2541 3081 3615 4157... |

CAUTION: Do not store data to a name that is routinely used by the TI-86. Such names are ANS, MATH, LOG, MODE, A, B, and so forth. Note that if you use a single letter as a name, this might cause one or more of the programs to not execute properly.

1.5.12 RECALLING STORED DATA The data you have stored remains in the memory of the TI-86 until you delete it using the instructions given in Section 1.5.15 of this *Guide*. When you wish to use the stored data, recall it to one of the lists L1, L2, ..., L5. We illustrate with the list named TAX, which we store in L2. Press [2nd] [−] (LIST) and under NAMES, find TAX. (You may need to press [MORE] until you see the list you want to recall.)

Press the F-key corresponding to the location of the list. Press [STO▸] 7 (L) [ALPHA] 2 [ENTER]. List L2 now contains the TAX data.

1.5.13 DELETING DATA You do not need to delete any data lists unless your memory is getting low or you just want to do it. For illustration purposes, we delete the TAX list.

Press [2nd] 3 (MEM) [F2] [DELET]. Next press [F4] [LIST] and use [▼] to move the cursor opposite TAX.
Press [ENTER]. To delete another list, use [▼] or [▲] to move the cursor opposite that list name and press [ENTER]. Exit this screen with [2nd] [EXIT] (QUIT) when finished.

WARNING: Be careful when in the DELETE menu. Once you delete something, it is gone from the TI-86 memory and cannot be recovered. If you mistakenly delete one of the lists L1, L2, ..., L5, you need to redo the statistical setup as indicated on page A-1 of this *Guide*.

1.5.14 WHAT IS "BEST FIT"? It is important to understand the method of least squares and the conditions necessary for its use if you intend to find the equation that best-fits a set of data. You can explore[4] the process of finding the line of best fit with program LSLINE. (Program LSLINE is given in the *TI-86 Program Appendix*.) For your investigations of the least-squares process with this program, it is better to use data that is not perfectly linear and data for which you do *not* know the best-fitting line.

Before using program LSLINE, clear all functions from the y(x)= list, turn on Plot1 by darkening the name Plot1 on the y(x)= screen, and enter your data in lists L1 and L2. (If Plot 1 is not turned on, the program will not execute properly.) Next, draw a scatter plot with [GRAPH] [F3] [ZOOM] [MORE] [F5] [ZDATA].

Press [F2] [WIND] and reset xScl and yScl so that you can use the tick marks to help identify points when you are asked to give the equation of a line to fit the data. Press [F5] [GRAPH] to draw the scatter plot and then return to the home screen.	

---

[4] This program works well with approximately 5 data points. Interesting data to use in this illustration are the height and weight, the arm span length and the distance from the floor to the navel, or the age of the oldest child and the number of years the children's parents have been married for 5 randomly selected persons.

TI-86 Guide

To activate program LSLINE, press [PRGM] [F1] [NAMES] followed by the F-key corresponding to the location of the program, and press [ENTER]. The program first displays the message shown to the right and pauses for you to read it.

NOTE: While the program is calculating, there is a small vertical line in the upper-right hand corner of the graphics screen that is dashed and "wiggly". This program pauses several times during execution for you to view the screen. Whenever the program pauses, the small line is "still" and you should press [ENTER] to resume execution after you have looked at the screen.

The program next shows the scatter plot of the data so that you can estimate the *y*-intercept and slope of *some* line that goes "through" the data. (You should not expect to guess the best-fit line on your first try!) Use the tick marks to estimate rise divided by run and note a possible *y*-intercept. Enter your guess for the slope and *y*-intercept after pressing [ENTER].

After pressing [ENTER] again, your line is drawn and the errors are shown as vertical line segments on the graph. (You may have to wait a moment to see the vertical line segments before again pressing [ENTER].)

Next the sum of squared errors, SSE, is displayed for your line. Decide whether you want to move the *y*-intercept of the line or change its slope to improve the fit to the data.

Press [ENTER] and enter 1 to choose the TRY AGAIN? option. After entering another guess for the *y*-intercept and/or slope, the process of viewing your line, the errors, and display of SSE is repeated.

If the new value of SSE is smaller than the SSE for your first guess, you have improved the fit.

When you feel an SSE value close to the minimum value is found, enter 2 at the TRY AGAIN? prompt. The program then overdraws the line of best fit on the graph and shows the errors for the line of best fit.

The program ends by displaying the coefficients *a* and *b* of the best-fit line $y = ax + b$ as well as the minimum SSE. Press [ENTER] to end the program. Use program LSLINE to explore[5] the method of least squares[6] that the TI-86 uses to find the line of best fit.

---

[5] Program LSLINE is for illustration purposes only. Actually finding the line of best fit for a set of data should be done according to the instructions in Section 1.5.7 of this *Guide*.

[6] Two articles that further explain "best-fit" are H. Skala, "Will the Real Best Fit Curve Please Stand Up?" Classroom Computer Capsule, *The College Mathematics Journal*, vol. 27, no. 3, May 1996 and Bradley Efron, "Computer-Intensive Methods in Statistical Regression," *SIAM Review*, vol. 30, no. 3, September 1988.

Chapter 2 Ingredients of Change: Nonlinear Models

2.1 Exponential and Logarithmic Functions and Models

As we begin to consider functions that are not linear, it is very important that you be able to draw scatter plots, find numerical changes in output data, and recognize the underlying shape of the basic functions to be able to identify which function best models a particular set of data. Finding the model is only a means to an end – being able to use mathematics to describe the changes that occur in real-world situations.

2.1.1 ENTERING EVENLY-SPACED INPUT VALUES

When an input list consists of many values that are the same distance apart, there is a calculator command that will generate the list so that you do not have to type the values in one by one. The syntax for this sequence command is *seq(formula, variable, first value, last value, increment)*. When entering years that differ by 1, the formula is the same as the variable and the increment is 1. Any letter can be used for the variable -- we choose to use x. We illustrate the sequence command use to enter the input for the population data in Table 2.2 in Section 2.1 of *Calculus Concepts*.

Year	1994	1995	1996	1997	1998	1999	2000	2001	2002	2003
Population	7290	6707	6170	5677	5223	4805	4420	4067	3741	3442

Clear any old data from lists L1 and L2. To enter the input data, position the cursor in the first list so that L1 is darkened. Generate the list of years beginning with 1994, ending with 2003, and differing by 1 with [2nd] [−] (LIST) [F5] [OPS] [MORE] [F3] [seq(] [x-VAR] [,] [x-VAR] [,] 1994 [,] 2003 [,] 1 [)].

Press [ENTER] and the sequence of years is pasted into L1. Manually enter the output values in L2.

2.1.2 FINDING PERCENTAGE DIFFERENCES

When the input values are evenly spaced, use program DIFF to compute the percentage differences (also called the *percentage changes*) in the output. If the data are perfectly exponential (that is, every data point falls on the exponential function), the percentage change in the output is constant. If the percentage changes are close to constant, this is an indication that an exponential model *may* be appropriate.

Return to the home screen. Find the percentage change in the population data with program DIFF. Press [PRGM] [F1] [NAMES] and then press the F-key that is under program DIFF.

Press [ENTER], choose [F1] [Yes] to continue, and at the Choice? prompt, press [F3] [%] to generate the list of percentage changes. Press [F4] [Quit] to exit the program.

The list will not scroll to the right because it was generated from within a program. However, the program has stored the percentage differences in list L5.

Press [2nd] [+] (STAT) [F2] [EDIT] ▶ ▶ ▶ to see the percentage changes in list L5. Scroll down L5 with ▼ to see all 9 values. The percentage differences are close to constant, so an exponential function will probably give a good fit to these data.

2.1.3 FINDING AN EXPONENTIAL FUNCTION TO MODEL DATA
Use your calculator to find an exponential equation to model the small town population data. The exponential function we use is of the form $y = ab^x$. It is very important that you align large numbers (such as years) whenever you find an exponential model. If you don't align to smaller values, the TI-86 may return an error message or an incorrect function. We continue to use the data in Table 2.2 of the text.

First, align the input data to smaller values. Other alignments are possible, but we choose to align the data so that $x = 0$ in 1994. To align, darken the list name and press [ALPHA] 7 (L) 1 [−] 1994 [ENTER].

Return to the home screen. Following the same procedure that you did to find a linear function, find the exponential function **and** paste the equation into the y1 location of the y(x)= list by pressing [2nd] [+] (STAT) [F1] [CALC] [F5] [ExpR] [ALPHA] 7 (L) 1 [,] [ALPHA] 7 (L) 2 [,] [2nd] [ALPHA] (alpha) 0 (y) 1.

Press [ENTER] [GRAPH] [F1] [y(x)=] to view y1. Be sure that Plot 1 is turned on, and press [EXIT] [F3] [ZOOM] [MORE] [F5] [ZDATA] to draw the scatter plot and the function graph.

As the percentage changes indicate, the function gives a very good fit for the data. To estimate the town's population in 2004, remember that x is the number of years past 1994 and evaluate the function in y1 at $x = 10$. We predict the population to be about 3167 people in 2004.

- Do not confuse the percentages changes (percentage differences) found from the data with the percentage change for the exponential function. The constant percentage change in the exponential function $y = ab^x$ is $(b − 1)100\%$. The constant percentage change for the function is a single value whereas the percentage changes calculated from the data are many different numbers.

2.1.4 FINDING A LOG FUNCTION TO MODEL DATA
Use your calculator to find a log equation of the form $y = a + b \ln x$. We illustrate finding this function with the air pressure and altitude data in Table 2.5 of Section 2.1 of *Calculus Concepts*.

Air pressure (inches of mercury)	13.76	5.56	2.14	0.82	0.33
Altitude (thousands of feet)	20	40	60	80	100

Clear the data that is currently in lists L1 and L2. Enter the air pressure data in L1 and the altitude data in L2. (If you wish, clear list L3, but it is not necessary to do so. Remember to clear the y(x)= list and turn Plot 1 on.)

A scatter plot of the data drawn with GRAPH F3 [ZOOM] MORE F5 [ZDATA] shows the slow decline that often indicates a log model.	
As when modeling linear and exponential functions, find and paste the log equation into the y1 location of the y(x)= list by pressing 2nd + (STAT) F1 [CALC] F4 [LnR] ALPHA 7 (L) 1 , ALPHA 7 (L) 2 , 2nd ALPHA (alpha) 0 (y) 1.	
Press y(x)= to view the function. Either press GRAPH F3 [ZOOM] MORE F5 [ZDATA] or F5 [GRAPH] to view the func-tion on the plot of the data. It appears to be a great fit.	

2.1.5 REVISITING INVERSE FUNCTIONS Recall that exponential and log functions are inverse functions. If you find the inverse function for the log equation that models altitude as a function of air pressure (given in the section above), it will be an exponential function. We briefly explore this concept using the TI-86. Leave the function $y = 76.174 - 21.331 \ln x$ in the y1 location the y(x)= list because we use it again in this investigation.

Find the exponential equation that models air pressure as a function of altitude and paste it into the y2 location by pressing 2nd + (STAT) F1 [CALC] F5 [ExpR] ALPHA 7 (L) 2 , ALPHA 7 (L) 1 , 2nd ALPHA (alpha) 0 (y) 2.	Note that L2 is typed first!
Press ENTER. Then press GRAPH F1 [y(x)=] and turn Plot 1 off. Enter the composite function y1 ∘ y2 (or y2 ∘ y1) in the y3 location. We use the Composition Property of Inverse Functions.	What is returned for y3 in the table is close to the value of x that was input. Press ▶ to see the y3 column.
You can also graphically verify that y1 and y2 are inverse func-tions. Delete or turn off the composite function(s) and turn off y2. On the home screen, have the TI-86 draw the graph of the inverse of y1 with GRAPH MORE F2 [Draw] MORE MORE MORE F3 [DrawInv] 2nd ALPHA (alpha) 0 (y) 1 ENTER. Press GRAPH F1 [y(x)=], turn y2 on, and press EXIT F5 [GRAPH].	You see the same graph, indicating that y2 and the inverse function are either the same or very close to the same function.

- There is a slight difference in the exponential function and the true inverse function for y1 because of rounding and because the exponential function in y2 was not found algebraically but from fitting an equation to data points.

2.2 Exponential Models in Action

In this section we consider further applications to equation solving with the TI-86 as well as its use in problems that deal with the mathematics of finance. You are probably familiar with the compound interest formulas. This section introduces you to some new methods of using your calculator with these familiar formulas.

2.2.1 NOMINAL AND EFFECTIVE INTEREST RATES
When you work with exponents and perform other calculations, it is very important to not round intermediate values. Values for nominal and effective interest rates can vary quite a bit if you do not use all available decimal places. We illustrate the proper procedure with the two investment option equations that are in part *a* of Example 1 in Section 2.2 of *Calculus Concepts*:

$$Q(t) = P\left(1 + \frac{0.069}{4}\right)^{4t} \quad \text{and} \quad M(t) = P\left(1 + \frac{0.067}{12}\right)^{12t}$$

The effective rates for each of the investment options given in part *a* of Example 1 are calculated as shown to the right. Either type in the entire expression or make use of the Ans memory.	`(1+.069/4)^4-1`
	`       .070805995356`
	`1+.067/12`
	`       1.00558333333`
	`Ans^12-1`
	`       .069096235282`

The effective rate for Q is about 7.1% and that for M is about 6.9%. Thus, the first option with quarterly interest has the higher effective rate and is therefore the better investment.

2.2.2 MORE ON SOLVING FOR INPUT VALUES
Solving for a variable that appears in an exponent is the same as solving for a variable that appears elsewhere in an equation when you use the equation solver. We illustrate this idea with the two investment option equations that are in part *b* of Example 1 in Section 2.2 of *Calculus Concepts*: $2P = Q(t)$ and $2P = M(t)$. The functions Q and M are defined in the previous section of this *Guide*.

Remember that when solving an equation on the TI-86, it must be entered in the calculator in the form $f(x) = constant$. If you prefer, you can divide both sides of the two equations by P. We choose to leave P in the equations to illustrate that it does not affect the solutions. We also choose to solve the equations $2P - Q(t) = 0$ and $2P - M(t) = 0$, but you can just as easily solve $Q(t) = 2P$ and $M(t) = 2P$ where the constant is $2P$ instead of 0.

Enter $2P - P\left(1 + \frac{0.069}{4}\right)^{4t}$ in y1 and $2P - P\left(1 + \frac{0.067}{12}\right)^{12t}$ in y2. Type P with ALPHA , and replace *t* with x. Be certain to enclose the exponents in parentheses as well as using the parentheses shown in the original formulas.	`Plot1 Plot2 Plot3`
	`\y1■2P-P(1+.069/4)^(...`
	`\y2■...+.067/12)^(12x)`
	
	`[x²] WIND ZOOM Trace Graph`
	`  x    y  INSf DELf SELCT `

Access the SOLVER with 2nd GRAPH (SOLVER) and clear any previous equations with ▲ CLEAR . To solve the equation y1 = 0, press F2 [y1] (or type y1) in the eqn: location and press ENTER . Enter 0 in the second exp= location.	`exp=y1`
	`exp=0■`
	`P=`
	`x=15.103`
	`bound=(-1E99,1E99)`
	
	`GRAPH WIND ZOOM TRACE SOLVE`

Enter some value for *P* and a guess for x. (Try using different values than those shown to the right to see what happens.) Have the cursor on the line containing the variable x and press F5 [SOLVE]. Repeat this procedure with a different value of *P* and a different guess for x to see that the same answer is obtained.

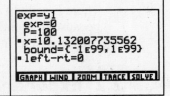

- Why do you obtain the same answer with a different value for *P* and a different guess? Go back to the original equations on the previous page to see that you can divide both sides by *P*, thereby eliminating it from the equations.

- What about the guess? Recall that the solver uses your guess as a starting point to find the answer. If there is more than one answer, the solver returns the solution closest to your guess. Here we are dealing with exponential functions on one side of the equation and a linear function (2*P*) on the other side. A line can cross an exponential curve only once, so there is only one answer for the solver to find.

To solve the second equation, press ▲ until you return to the eqn: screen, change y1 to y2, enter a new guess (or leave what is currently there as the guess), be sure the cursor is on the line corresponding to what you are solving for (x), and press F5 [SOLVE].

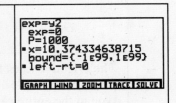

CAUTION: Being able to solve equations is only the first step in obtaining a solution. Always read again the question you are asked and see what you must do with the result from the solver to answer the question. You must also determine if the answer should be rounded due to a discrete interpretation. Your calculator cannot assist you with these tasks, but practice helps a lot!

2.2.3 FINDING FORMULA OUTPUTS BY USING PREVIOUS ENTRIES

You can enter a formula in the y(x)= list and find outputs by either by evaluating the formula at different values on the home screen or by using the ASK feature of the TI-86 TABLE. (See Sections 1.1.5 and 1.1.6.) Another method that you can use is to recall expressions previously typed by repeatedly using the calculator's last entry feature. Often, this method can save time. We illustrate it by using the values in Table 2.23 in Section 2.2 of *Calculus Concepts*.

On the home screen, type the formula $\left(1+\frac{1}{n}\right)^n$. Type *n*, so that it prints as a capital letter on the calculator, with ALPHA 9. Press ENTER. Next, store 1 in *n*.

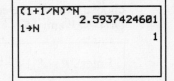

NOTE: The output you obtain depends on the value that is stored for N. You may obtain a different output value than the one shown above because you have a different value stored in N. After storing 1 in N, your next result should be the same as the shown below.

Press 2nd ENTER (ENTRY) twice (or as many times as it takes to again display the formula on the screen), and then press ENTER. The formula is now evaluated at *n* = 1. Store 2 in *n* and repeat the process of recalling and evaluating the formula.

Continue in this manner until you find all the amount values in Table 2.23 of the text.	`(1+1/N)^N` 525600
	              2.7182792154
	`31536000→N`
	                 31536000
	`(1+1/N)^N`
	              2.71828247254

2.2.4 DETERMINING FUTURE VALUE
You can often save many keystrokes with the recall of previously-typed expressions by repeatedly using the calculator's last entry feature. When a formula contains more than one input variable, it is easier to recall the last entry on the home screen than to try to use the TI-86 table.

To illustrate, consider the compound interest formula – one that contains several input variables: Amount = $P\left(1+\frac{r}{n}\right)^{nt}$ dollars is the amount in an account paying $100r$% interest (compounded n times a year) on an initial deposit of P over a period of t years.

To determine the amount in an account with $P = \$1500$, $r = 0.075$, $n = 12$, and $t = 1$, first store these values in the memory locations P, R, N, and T.	`1500→P` 1500 `.075→R` .075 `12→N` 12 `1→T` 1
Also on the home screen, type in the formula for the amount in the account – the future value. Remember to use parentheses around the exponent because it consists of more than one symbol. Press ENTER to evaluate the formula at the stored values.	`12→N` .075 `1→T` 12 `P(1+R/N)^(N*T)` 1 1616.44889828

WARNING: If you forget to put the times sign between N and T in the future value formula, you will probably obtain an UNDEFINED error because the TI-86 considers NT as the name of a new variable.

To determine the future value of the account in 2 years, first store 2 in t. Press 2nd ENTER (ENTRY) twice (or as many times as it takes to again display the amount formula on the screen), and then press ENTER. (The other values remain as you stored them until they are changed.)	`2→T` 2 `P(1+R/N)^(N*T)` 1741.93802718

2.2.5 FINDING PRESENT VALUE
Suppose you want an alternate method than that indicated in the text to solve the equation in part *a* of Example 5 in Section 2.2 of *Calculus Concepts*:

Solve $9438.40 = P\left(1+\frac{0.075}{12}\right)^{60}$ for the present value P.

Access the SOLVER with 2nd GRAPH (SOLVER) and clear any previous equations with CLEAR. To solve this equation, first enter the side containing the variable P in the eqn: location.	`eqn:P(1+.075/12)^60`

Press ENTER. Enter the constant 9438.40 on the second line. Enter a guess for the unknown quantity, P. With the cursor on the line containing the variable P, press F5 [SOLVE].	`exp=P(1+.075/12)^60`
	`exp=9438.4`
	`■P=6494.4858703433`
	`bound={-1E99,1E99}`
	`■left-rt=0`

2.3 Logistic Functions and Models

This section introduces the logistic function that can be used to describe growth that begins as exponential and then slows down to approach a limiting value. A change in the sign of one of the parameters gives a logistic function that describes decay instead of growth. Logistic functions are often in applications to the spread of disease. The TI-86's random number generator can be used to collect data to illustrate such a situation.

2.3.1 GENERATING RANDOM NUMBERS Imagine all the real numbers between 0 and 1, including 0 but not 1, written on identical slips of paper and placed in a hat. Close your eyes and draw one slip of paper from the hat. You have just chosen a number "at random."

Your calculator doesn't offer you a random choice of all real numbers between 0 and 1, but it allows you to choose, *with an equal chance of obtaining each one*, any of 10^{12} different numbers between 0 and 1 with its random number generator called rand. Before using your calculator's random number generator for the first time, you should "seed" the random number generator. (This is like mixing up all the slips of paper in the hat.)

Pick some *whole* number, <u>not</u> the one shown on the right, and store it as a "seed" with *your number* [STO▶] [2nd] [×] (MATH) [F2] [PROB] [F4] [rand] [ENTER]. Everyone needs to have a different seed, or the number choices will not be random.

Enter rand again, and press [ENTER] several times.

Your list of random numbers should be different from the one on the right. Notice that all numbers are between 0 and 1.

We usually count with the numbers 1, 2, 3, ... and so forth, not with decimal values. So, the TI-86 has a key called randInt that chooses a whole number between, and including, 1 and N.

Enter, using a specific number for N, [F5] [randIn] [(] 1 [,] N [,] 1 [)]. For instance, if you want to choose a person at random from a room that contains 10 people, use $N = 10$.

Repeatedly press [ENTER] to choose more random numbers.

As, previously stated, your list of numbers should be different from those shown above. All values will be between 1 and 10. Note that it is possible to obtain the same value more than once. This corresponds to each slip of paper being put back in the hat after being chosen and the numbers mixed well before the next number is drawn.

NOTE: If you want to generate a list of 5 random numbers, change the last 1 in the above instruction to 5; if you want 10 random numbers, change it to 10, and so forth.

2.3.2 FINDING A LOGISTIC FUNCTION TO MODEL DATA You can use your calculator to find a logistic function of the form $y = \dfrac{L}{1 + Ae^{-Bx}} + d$. Unlike the other functions we have fit to data, the logistic equations that you obtain will be slightly different from those given in the text answer key and different from logistic equations found with another model calculator because the TI-86 equation has a built-in vertical shift (d). (The logistic equations in *Calculus Concepts* were found using a TI-83 that does not have a built-in vertical shift.) As you did when finding an exponential equation for data, large input values must be aligned or an error or possibly an incorrect answer could be the result.

Note that the TI-86 finds a "best-fit" logistic function rather than a logistic function with a limiting value L such that no data value is ever greater than L. We illustrate finding a logistic function using the data in Example 1 of Section 2.3 in *Calculus Concepts*.

End of month	Jan	Feb	Mar	Apr	May	June	July	Aug	Sept
Total number of swimsuits	4	12	25	58	230	439	648	748	769

Clear any old data. Delete any functions in the y(x)= list and turn on Plot 1. Using an input of 1 for January, 2 for February, and so forth, enter the data in the above table in lists L1 and L2:

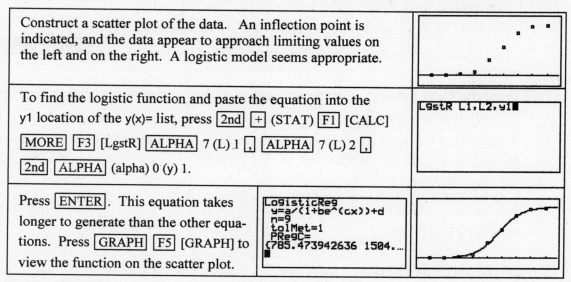

Construct a scatter plot of the data. An inflection point is indicated, and the data appear to approach limiting values on the left and on the right. A logistic model seems appropriate.

To find the logistic function and paste the equation into the y1 location of the y(x)= list, press [2nd] [+] (STAT) [F1] [CALC] [MORE] [F3] [LgstR] [ALPHA] 7 (L) 1 [,] [ALPHA] 7 (L) 2 [,] [2nd] [ALPHA] (alpha) 0 (y) 1.

Press [ENTER]. This equation takes longer to generate than the other equations. Press [GRAPH] [F5] [GRAPH] to view the function on the scatter plot.

NOTE: The TI-86 uses the variable a for the value that the text calls L, b for the value the text calls A, and c for the value the text calls B. The variable d represents the vertical shift. You can see the values of these parameters by pressing [▶] to view the list {a b c d} that is shown in the middle box above or you can press [▶] and see them in the equation in the y(x)= list.

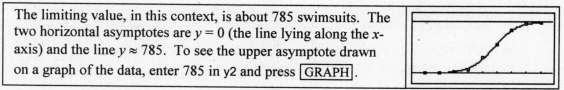

The limiting value, in this context, is about 785 swimsuits. The two horizontal asymptotes are $y = 0$ (the line lying along the x-axis) and the line $y \approx 785$. To see the upper asymptote drawn on a graph of the data, enter 785 in y2 and press [GRAPH].

CAUTION: The limiting values as given above can be interpreted in the real-life situation. Without considering the context of the problem, the two horizontal asymptotes according to the logistic function found on the TI-86 are $y \approx 0.88$ and $y \approx 785.5$ (Note that $d \approx 0.887$ and we round down in and out of the problem context so that the curve does not go below the asymptote.)

- Provided the input values are evenly spaced, program DIFF might be helpful when you are trying to determine if a logistic model is appropriate for certain data. If the first differences begin small, peak in the middle, and end small, this is an indication that a logistic model may provide a good fit to the data. Such is true for the swimsuit data because the first differences are 8, 13, 33, 172, 209, 209, 100, and 21.

2.3.3 RECALLING MODEL PARAMETERS
Rounding function parameters can often lead to incorrect or misleading results. You may find that you need to use the complete values of the coefficients after you have found a function that best fits a set of data. It would be tedious to copy all these digits in a long decimal number into another location of your calculator. You don't have to! The following procedure applies for any function that you find using one of the built-in regressions (*i.e.*, a function from the STAT CALC menu) in your TI-86.

We illustrate these ideas by recalling the parameters for the logistic function found in the previous section. These parameters can be recalled as a list or as individual numbers.

To recall the entire list, which contains the coefficients listed in alphabetical order, press [2nd] [+] (STAT) [F5] [VARS] [MORE] [MORE] [MORE] [MORE] [F2] [PRegC] [ENTER]. Recall that ▶ scrolls the list to the right for viewing all the values.

Using the TI-86 Catalog: You can obtain most of the calculator commands by pressing [2nd] [CUSTOM] [CATLG-VARS], choosing the appropriate F-key (use [F1] [CATLG] or [F2] [ALL] if you are not sure what type of item you are trying to find), pressing the key corresponding to the first letter of the name of the item, using ▼ until the cursor is opposite the item, and then pressing [ENTER] to copy the item to wherever you had the cursor before entering the catalog. For example, to type PRegC, press [2nd] [CUSTOM] [CATLG-VARS] [MORE] [MORE] [F4] [STAT] ▼ [ENTER].

To recall the value of a in $y = a/(1+ be^{(-cx)}) + d$, first follow the directions given above to type PRegC on the home screen. Then, because a is first in the list, press [(] 1 [)] [ENTER]. To recall b, change 1 to 2 because b is second in the list, and so forth.

- This same procedure applies to any equation you have found using the STAT CALC menu. Observe the name of the list that appears when you find the equation and call up that list to find the coefficients in the function. If no list appears (such as when finding a linear or exponential function), the coefficients are given as a and b in the CATLG STAT menu.

WARNING: Once a different function is found, no parameters from a previous function that you have fit to data remain stored in the calculator's memory.

2.4 Polynomial Functions and Models
You will in this section learn how to fit functions that have the familiar shape of a parabola or a cubic to data. Using your calculator to find these equations involves the same procedure as when using it to find linear, exponential, log, or logistic functions.

2.4.1 FINDING SECOND DIFFERENCES
When the input values are evenly spaced, you can use program DIFF to quickly compute second differences in the output values. If the data are perfectly quadratic (that is, every data point falls on a quadratic function), the second differences in the output values are constant. When the second differences are close to constant, a quadratic function *may* be appropriate for the data.

We illustrate these ideas with the roofing jobs data given at the beginning of Section 2.4:

TI-86 Guide

Month of the year	1	2	3	4	5	6
Number of roofing jobs	12	14	22	37	58	84

We align the input data so that 1 = January, 2 = February, etc. Clear any old data and enter the data in the above table in lists L1 and L2. The input values are evenly spaced, so we can see what information is given by viewing the second differences.

Run program DIFF and find the first differences (also in list L3), the second differences (also in L4), and the percentage changes (also in list L5). The second differences are close to constant, so a quadratic function may give a good fit for these data.	Choice? 1st differences in L3 {2 8 15 21 26} 2nd differences in L4 {6 7 6 5} percent change in L5 {16.6666666667 57.14... [1st] [2nd] [%] [Quit]
Construct a scatter plot of the data. (Don't forget to clear the y(x)= list of previously-used equations and to turn on Plot 1.) The shape of the data confirms the numerical investigation result that a quadratic function is appropriate.	

2.4.2 FINDING A QUADRATIC FUNCTION TO MODEL DATA

Use your TI-86 to find a quadratic function to model the roofing jobs data. The TI-86's quadratic function is of the form $y = ax^2 + bx + c$ and is fit with the command P2Reg.

Find the quadratic function and paste the equation into y1 with [2nd] [+] (STAT) [F1] [CALC] [MORE] [F4] [P2Reg] [ALPHA] 7 (L) 1 [,] [ALPHA] 7 (L) 2 [,] [2nd] [ALPHA] (alpha) 0 (y) 1.	P2Reg L1,L2,y1
Press [ENTER] and then use [GRAPH] [F5] [GRAPH] to overdraw the parabola on the scatter plot.	QuadraticReg y=ax²+bx+c n=6 PRegC= {3.07142857143 -7.01...
How many jobs do we predict the company will have in August? Because August corresponds to $x = 8$, evaluate y1 at $x = 8$. We predict that there will be 156 jobs in August.	y1(8) 156.257142857

2.4.3 FINDING A CUBIC FUNCTION TO MODEL DATA

Whenever a scatter plot of data shows a single change in concavity, we are limited to fitting either a cubic or logistic function. If one or two limiting values are apparent, use the logistic equation. Otherwise, a cubic function should be considered. The TI-86's cubic function is of the form $y = ax^3 + bx^2 + cx + d$ and is accessed using the P3Reg key.

We illustrate finding a cubic function with the data in Table 2.44 in Example 3 of Section 2.4 in *Calculus Concepts*. The data give the average price in dollars per 1000 cubic feet of natural gas for residential use in the U.S. for selected years between 1980 and 1997.

Year	1980	1982	1985	1990	1995	1996	1997
Price (dollars)	3.68	5.17	6.12	5.77	6.06	6.34	6.93

First, clear your lists, and then enter the data. Next, align the input data so that x represents the number of years since 1980. (We do not have to align here, but we do so in order to have smaller coefficients in the cubic function.)	
Draw a scatter plot of these data. Notice that a concavity change is evident, but there do not appear to be any limiting values. Thus, a cubic model is appropriate to fit the data.	
Find the cubic function and paste the equation into y1 with $\boxed{2nd}$ $\boxed{+}$ (STAT) $\boxed{F1}$ [CALC] $\boxed{MORE}$ $\boxed{F5}$ [P3Reg] $\boxed{ALPHA}$ 7 (L) 1 $\boxed{,}$ $\boxed{ALPHA}$ 7 (L) 2 $\boxed{,}$ $\boxed{2nd}$ $\boxed{ALPHA}$ (alpha) 0 (y) 1.	
Press $\boxed{ENTER}$ and then use $\boxed{GRAPH}$ $\boxed{F5}$ [GRAPH] to overdraw the cubic function on the scatter plot.	
What was the price of 1000 cubic feet of natural gas in 1994? Because x represents the number of years after 1980, evaluate y1 at $x = 14$ to find that the answer is $5.77.	

According to the model, when did the price of 1000 cubic feet of natural gas first exceed $6? You can find the answer by using the SOLVER to find the solution to y1 = 6. However, since we already have a graph of this function, the following may be easier. Note that if you do use the SOLVER, you should first TRACE the graph to estimate a guess for x that you need to enter in the SOLVER.

Press $\boxed{GRAPH}$ $\boxed{F1}$ [y(x)=] and enter 6 in y2. Press $\boxed{EXIT}$ $\boxed{F5}$ [GRAPH]. Note that there are 3 solutions to the equation y1 = 6. We are asked to find the one that occurs first. Press $\boxed{MORE}$ $\boxed{F1}$ [MATH] $\boxed{MORE}$ $\boxed{F3}$ [ISECT], press $\boxed{ENTER}$ $\boxed{ENTER}$ to choose the 2 functions, and use $\boxed{\blacktriangleleft}$ to move the cursor near the first intersection point.	
Press $\boxed{ENTER}$ to find the input at the point of intersection. Because x is the number of years after 1980, the answer to the question posed in part c of Example 3 is either 1984 or 1985.	
The price had not exceeded $6 in 1984, so the answer to the question is 1985. Carefully read the question you are asked!	

2.5 Shifting Data

If a scatter plot of data, numerical differences calculated from data, and/or a particular context indicate that a certain function is appropriate but the function does not appear to fit the data, it may be that a horizontal or vertical shift of the data should be considered.

2.5.1 SHIFTING DATA VERTICALLY Because the TI-86 has a built-in vertical shift for the logistic function, this section of the *Guide* is not necessary for logistic models. Thus, instead of the investment club data in Table 2.66 in Example 1 of Section 2.5 of *Calculus Concepts* (which uses a logistic function), we use the exponential data in Table 2.64 on page 155 of the text to see if a vertical shift improves the fit.

x	0	2	4	6	8
f(x)	101	104	116	164	356

First, clear your lists, and then enter the data in the above table in L1 and L2.

Draw a scatter plot of the data, which indicates that an exponential model may be appropriate.	
However, when we find an exponential function, it is a poor fit for the data. Also notice how the scatter plot appears to level off as the input approaches zero.	

This situation calls for shifting the data vertically so that it approaches a lower asymptote of $y = 0$, which is what the exponential function in the TI-86 has as its lower asymptote for an increasing exponential curve. There are many choices of how to shift the data. We choose to subtract 100 from each output data value.

Highlight L2 at the top of the second column, type L2 – 100, and then press [ENTER]. Now fit an exponential function to the shifted data.	
The shifted function is therefore $g(x) = 2x + 100$. Add 100 to the function in y1.	
If you wish to graph g on the original data, add 100 to each output value by highlighting L2 at the top of the second column, typing L2 – 100, and then pressing [ENTER].	
Draw the graphs of g and the original data with [GRAPH] [F3] [ZOOM] [MORE] [F5] [ZDATA].	

Chapter 3   Describing Change: Rates

3.1 Change, Percent Change, and Average Rates of Change

As you calculate average and other rates of change, remember that every numerical answer in a context should be accompanied by units telling how the quantity is measured. You should also be able to interpret each numerical answer. It is only through their interpretations that the results of your calculations will be useful in real-world situations.

3.1.1 FINDING AVERAGE RATE OF CHANGE
Finding an average rate of change using a function is just a matter of evaluating the equation at two different values of the input variable and dividing by the difference of those input values.

We illustrate this concept using the function describing the population density of Nevada from 1950 through 1997 that is given in Example 3 of Section 3.1 of *Calculus Concepts*.

The population density referred to above is given by the function $p(t) = 0.1617(1.04814^t)$ people per square mile where t is the number of years after 1900. Press GRAPH F1 [y(x)=], clear any functions, turn Plot 1 off, and enter this function in y1 using x as the input variable.	

To find the average rate of change of the population density from 1950 through 1980, first realize that 1950 corresponds to $x = 50$ and 1980 corresponds to $x = 80$. Then, recall that the average rate of change of p between 1950 and 1980 is given by the quotient $\frac{p(80)-p(50)}{80-50}$.

Return to the home screen and type the quotient $\frac{y1(80)-y1(50)}{80-50}$. Remember to enclose both the numerator and denominator of the fraction in parentheses. Finding this quotient in a single step avoids having to round intermediate calculation results.	(y1(80)-y1(50))/(80-5 0) .175244367956

Recall that rate of change units are output units per input units. On average, Nevada's population density increased about 0.18 people per square mile per year between 1950 and 1980.

To find the average rate of change between 1980 and 1997, recall the last expression with 2nd ENTER (ENTRY) and type over 50 with 97 in two places. Press ENTER.	(y1(80)-y1(50))/(80-5 0)
	   .175244367956
	(y1(80)-y1(97))/(80-9 7)
	   .500695677834

- If you have many average rates of change to calculate, you could put the average rate of change formula in the graphing list: y2 = (y1(B) − y1(A))/(B − A). Of course, you need to have a function in y1. Then, on the home screen, store the inputs of the two points in *A* and *B* with 80 STO▶ A 2nd . (:) ALPHA 90 STO▶ B ENTER. All you need do next is type y2 and press ENTER. Then, store the next set of inputs into *A* and *B* and recall y2, using 2nd ENTER (ENTRY) to recall each instruction, press ENTER, and you have the average rate of change between the two new points. Try it!

3.1.2 CALCULATING PERCENTAGE CHANGE
You can find percentage changes using data either by the formula or by using program DIFF. To find a percentage change from a function

instead of data, you should use the percentage change formula. We again use the population density of Nevada function in Example 3 of Section 3.1 of *Calculus Concepts* to illustrate.

| Have $p(x) = 0.1617(1.04814^x)$ in y1. The percentage change in the population density between 1950 and 1997 is given by the formula $\frac{p(97)-p(50)}{p(50)} \cdot 100\% = \frac{y1(97)-y1(50)}{y1(50)} \cdot 100\%$. Recall that if you type in the entire quotient, you avoid rounding errors. | (y1(97)-y1(50))/y1(50
)
              8.11395489512
Ans*100
              811.395489512 |

The population density of Nevada increased by about 811.4% between 1950 and 1997.

3.2 Instantaneous Rates of Change

We first examine the principle of local linearity, which says that if you are close enough, the tangent line and the curve are indistinguishable. We also explore two methods for using the calculator to draw a tangent line at a point on a curve.

3.2.1 MAGNIFYING A PORTION OF A GRAPH
The ZOOM menu of your calculator allows you to magnify any portion of a graph. Consider the graph shown in Figure 3.7 in Section 3.2 of *Calculus Concepts*. The temperature model is $T(x) = -0.804x^2 + 11.644x + 38.114$ degrees Fahrenheit where x is the number of hours after 6 a.m.

Enter the temperature equation in y1 and draw the graph between 6 a.m. ($x = 0$) and 6 p.m. ($x = 12$) using [GRAPH] [F3] [ZOOM] [MORE] [F1] [ZFIT]. We now want to zoom and "box" in several points on the graph to see a magnified view.	
The first point we consider on the graph is point A with $x = 3$. Press [F3] [ZOOM] [F1] [Box] and use the arrow keys ([◄], [▲], etc.) to move the cursor to the left of the curve close to where $x = 3$. (You may not have the same coordinates as those shown on the right.) Press [ENTER] to fix the lower left corner of the box.	x=2.8571428571 y=67.353169629
Use the arrow keys to move the cursor to the opposite corner of your "zoom" box. Point A should be close to the center of your box.	x=4 y=70.073092385
Press [ENTER] to magnify the portion of the graph that is inside the box. Look at the dimensions of the view you now see with [EXIT] [WIND]. Repeat the above process if necessary. The closer you zoom in on the graph, the more it looks like a line.	

Reset xMin to 0, xMax to 12, and redraw the graph of T with ZFIT. Repeat the zoom and box steps for point B with $x = 7.24$ and point C with $x = 9$. Note for each point how the graph of the parabola looks more like a line the more you magnify the view. These are illustrations of *local linearity*.

3.2.2 DRAWING A TANGENT LINE

The TI-86 can draw a line tangent to the graph of a function at a specified point. To illustrate the process, we draw several tangent lines on the graph of T, the temperature function given in the previous section of this *Guide*.

Enter $T(x) = -0.804x^2 + 11.644x + 38.114$ in y1. Next, press F2 [WIND] and set xMin = 0, xMax = 12, yMin = 38, and yMax = 85. Press F5 [GRAPH]. On the graph screen, press MORE F1 [MATH] MORE MORE F1 [TANLN] 3 ENTER, and the tangent line is drawn at $x = 3$.

Note that the slope of the tangent line, symbolized by dy/dx, is displayed at the bottom of the screen. Compare this line with the zoomed-in view of the curve at point A (in the previous section). They actually are the same line if you are at point A!

We illustrate a slightly different method to draw the tangent line to the curve at point B where $x = 7.24$. Press GRAPH. Then press MORE F2 [DRAW] MORE MORE MORE F2 [TanLn(] 2nd ALPHA (alpha) 0 (y) 1 , 7.24).

Press ENTER and the tangent line is drawn at the high point of the parabola. Compare this line with the zoomed-in view of the curve at point B (in the previous section).

Note that this method of drawing the tangent does not show the slope of the tangent line.

Choose the method you prefer from the two methods given above and draw the tangent at point C where $x = 9$.

(Compare your graph with the one shown in Figure 3.9 in Section 3.2 of the text.)

3.2.3 VISUALIZING THE LIMITING PROCESS

This section of the *Guide* is optional, but it might help you understand what it means when we say that the tangent line is the limiting position of secant lines. Program SECTAN can be used to view secant lines between a point $(a, f(a))$ and close points on a curve $y = f(x)$. The program also draws the tangent line at the point $(a, f(a))$. Before using program SECTAN, a function (using x as the input variable) must be entered in the y1 location of the y(x)= list, and you must draw a graph of the function. We illustrate with the graph of the function T that was used in the previous two sections, but you can use this program with any graph. (Program SECTAN is in the *TI-86 Program Appendix*.)

Caution: In order to properly view the secant lines and the tangent line, it is essential that you first draw a graph of the function clearly showing the function, the point of tangency (which should be near the center of the graph), and enough space so that the close points on either side can be seen.

Have $T(x) = -0.804x^2 + 11.644x + 38.114$ in y1. Next, press F2 [WIND] and set xMin = 0, xMax = 7, yMin = 30, and yMax = 90. Press F5 [GRAPH]. Press PRGM F1 [NAMES] and the F-key corresponding to the location of program SECTAN.

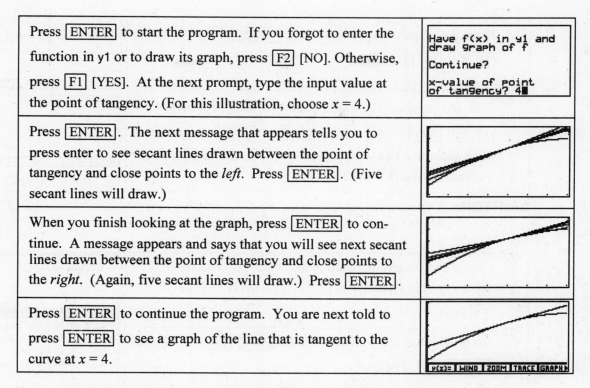

Press ENTER to start the program. If you forgot to enter the function in y1 or to draw its graph, press F2 [NO]. Otherwise, press F1 [YES]. At the next prompt, type the input value at the point of tangency. (For this illustration, choose $x = 4$.)	
Press ENTER. The next message that appears tells you to press enter to see secant lines drawn between the point of tangency and close points to the *left*. Press ENTER. (Five secant lines will draw.)	
When you finish looking at the graph, press ENTER to continue. A message appears and says that you will see next secant lines drawn between the point of tangency and close points to the *right*. (Again, five secant lines will draw.) Press ENTER.	
Press ENTER to continue the program. You are next told to press ENTER to see a graph of the line that is tangent to the curve at $x = 4$.	

As you watch the graphs, you should notice that the secant lines are becoming closer and closer to the tangent line as the close point moves closer and closer to the point of tangency.

3.2.4 TANGENT LINES AND INSTANTANEOUS RATES OF CHANGE

Sometimes the TI-86 gives results that are not the same as the mathematical results you expect. This does not mean that the calculator is incorrect – it does, however, mean that the calculator programming is using a different formula or definition than the one that you are using. You need to know when the TI-86 produces a different result from what you expect. The results the TI-86 gives involving instantaneous rates of change may vary depending on which derivative is chosen in the MODE screen. For the following discussion, we assume that you have dxDer1 chosen in the last row of the MODE screen as indicated in Figure 1 on page B-1.

In this section, we investigate what the calculator does if you ask it to draw a tangent line where the line cannot be drawn. Consider these special cases:

1. What happens if the tangent line is vertical? We consider the function $f(x) = (x + 1)^{1/3}$ which has a vertical tangent at $x = -1$.

2. How does the calculator respond when the instantaneous rate of change at a point does not exist? We illustrate with $g(x) = |x| - 1$, a function that has a sharp point at $(0, -1)$.

3. Does the calculator draw the tangent line at the break point(s) of a piecewise continuous function? We consider two situations:
   a. $h(x)$, a piecewise continuous function that is continuous at all points, and
   b. $m(x)$, a piecewise continuous function that is not continuous at $x = 1$.

1. Enter the function $f(x) = (x + 1)^{1/3}$ in the y1 location of the y(x)= list. Remember that anytime there is more than one symbol in an exponent and you are not sure of the TI-86's order of operations, enclose the power in parentheses.	

Draw the graph of the function with EXIT F3 [ZOOM] MORE F4 [ZDECM].	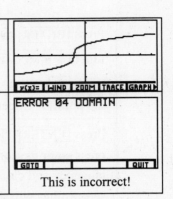
With the graph on the screen, press MORE F1 [MATH] MORE MORE F1 [TANLN] (−) 1 ENTER. Instead of the tangent line being drawn at $x = -1$, you see an error message. Press F5 [QUIT].	This is incorrect!

CAUTION: A vertical tangent line at $x = -1$ does not draw (as it should). The slope of the tangent line does not exist, but the tangent line can be drawn.

2. Clear y1 and enter the function $g(x) = |x| - 1$. The absolute value symbol is typed with 2nd X (MATH) F1 [NUM] F5 [abs].	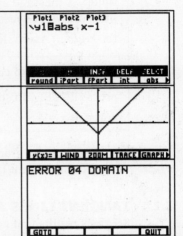
Draw the graph of the function with EXIT F3 [ZOOM] MORE F4 [ZDECM].	
With the graph on the screen, press MORE F1 [MATH] MORE MORE F1 [TANLN] 0 ENTER. You see an error message. Press F5 [QUIT].	

This is correct! There is a sharp point on the graph of g at $(0, -1)$, and the limiting positions of secant lines from the left and the right at that point are different. A tangent line cannot be drawn on the graph of g at $(0, -1)$ according to our definition of instantaneous rate of change.

3a. Clear y1 and y2 and enter, as indicated, the function $$h(x) = \begin{cases} x^2 & \text{when } x \leq 1 \\ x & \text{when } x > 1 \end{cases}$$ Recall that the inequality symbol menu is accessed with 2nd 2 (TEST).	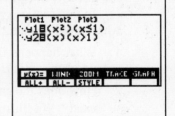

Set each part of the function to draw in DOT mode by pressing MORE F3 [STYLE] and continuing to press F3 until the slanted line turns to a dotted line.

Draw the graph of the function with EXIT F3 [ZOOM] MORE F4 [ZDECM].	

With the graph on the screen, press MORE F1 [MATH] MORE MORE F1 [TANLN] 1 ENTER. You see an error message. Press F5 [QUIT].

This is correct! Even though h is continuous for all values of x, there is no tangent line to the graph of h at the break point $x = 1$. Secant lines drawn using close points on the right and on the left of $x = 1$ do not approach the same slope, so the instantaneous rate of change of h does not exist at $x = 1$.

NOTE: There is <u>no tangent line at the break point</u> of a piecewise continuous function (even if that function is continuous at the break point) unless secant lines drawn through close points to the left and right of that point approach the same value.

3b. Edit y2 by placing the cursor over the right parenthesis following x and pressing 2nd DEL (INS) + 1 to enter

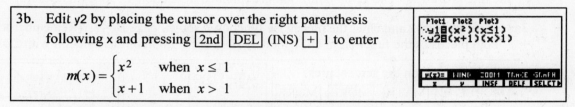

$$m(x) = \begin{cases} x^2 & \text{when } x \leq 1 \\ x+1 & \text{when } x > 1 \end{cases}$$

The TI-86 should still be in DOT mode from the previous graph.

Draw the graph of the function with EXIT F3 [ZOOM] MORE F4 [ZDECM].

With the graph on the screen, press MORE F1 [MATH] MORE MORE F1 [TANLN] 1 ENTER. You see an error message. Press F5 [QUIT].

This is correct! Because $m(x)$ is not continuous at $x = 1$, the instantaneous rate of change does not exist at that point. The tangent line cannot be drawn on the graph of m when $x = 1$.

The two methods that the TI-86 uses to compute instantaneous rates of change give the same numerical results for smooth continuous functions. However, the results for the two methods are not the same for all functions. The method the TI-86 uses to compute instantaneous rates of change with each of the two methods is explained in detail in Section 4.3.1 of this *Guide*. We now see what results are given for the previous investigations when the calculator's mode screen is set to dxNDer instead of dxDer1. Press 2nd MORE (MODE) and choose dxNDer in the last row of the MODE screen.

1. Draw the graph of $f(x) = (x + 1)^{1/3}$ with EXIT F3 [ZOOM] MORE F4 [ZDECM].

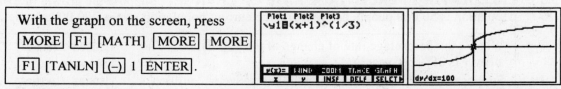

With the graph on the screen, press MORE F1 [MATH] MORE MORE F1 [TANLN] (–) 1 ENTER.

CAUTION: Note that the TI-86 correctly draws the vertical tangent. However, the TI-86 says the slope is 100, but the slope of this line at (–1, 0) does not exist because the tangent is vertical.

2. Clear y1 and enter the function $g(x) = |x| - 1$. (Type the absolute value symbol with [2nd] [X] (MATH) [F1] [NUM] [F5] [abs].) Draw the graph of the function with ZDECM.

With the graph on the screen, press [MORE] [F1] [MATH] [MORE] [MORE] [F1] [TANLN] 0 [ENTER].

THIS IS INCORRECT! There is a sharp point on the graph of g at $(0, {}^-1)$, and the limiting positions of secant lines from the left and the right at that point are different. A tangent line cannot be drawn on the graph of g at $(0, {}^-1)$ according to *our* definition of instantaneous rate of change. The TI-86's definition using the dxNDer mode is different, and this is why the line is drawn. (The TI-86 definition is explained in Section 4.3.1 of this *Guide*.)

3a. Clear y1 and y2 and enter the piecewise function h shown page B–47. (Remember to put both pieces of the function in DOT mode.) Draw the graph of the function with ZDECM.

With the graph on the screen, press [MORE] [F1] [MATH] [MORE] [MORE] [F1] [TANLN] 1 [ENTER].

THIS IS INCORRECT! Even though h is continuous for all values of x, there is no tangent to the graph of h at $x = 1$ and the instantaneous rate of change of h does not exist at $x = 1$.

3b. Clear y1 and y2 and enter the piecewise function m shown page B–48. (Remember to put both pieces of the function in DOT mode.) Draw the graph of the function with ZDECM.

With the graph on the screen, press [MORE] [F1] [MATH] [MORE] [MORE] [F1] [TANLN] 1 [ENTER].

THIS IS INCORRECT! Because m is not continuous at $x = 1$, the instantaneous rate of change does not exist at that point. Press [2nd] [MORE] (MODE) and choose dxDer1.

CAUTION: Be certain that the instantaneous rate of change exists at a point and the tangent line can be drawn at that point before using your calculator to draw a tangent line or use the slope of the tangent line that may be printed on the screen.

3.3 Derivatives

There are no new calculator techniques in this section, but we illustrate a new calculation.

3.3.1 CALCULATING PERCENTAGE RATE OF CHANGE
Suppose the growth rate of a population is 50,000 people per year and the current population size is 200,000 people.

What is the percentage rate of change of the population? The answer is 25% per year.

Suppose instead that the current population size is 2 million. What is the percentage rate of change? The answer is 2.5% per year, which is a much smaller percentage rate of change.

```
50000/200000
              .25
Ans*100
               25
50000/(2*10^6)
             .025
Ans*100
              2.5
```

Chapter 4 Determining Change: Derivatives

4.1 Numerically Finding Slopes

Using your calculator to find slopes of tangent lines does not involve a new procedure. However, the techniques that are discussed in this section allow you to repeatedly apply a method of finding slopes that gives quick and accurate results.

4.1.1 NUMERICALLY ESTIMATING SLOPES ON THE HOME SCREEN   Finding the slopes of secant lines joining the point at which the tangent line is drawn to increasingly close points on a function to the left and right of the point of tangency is easily done using your TI-86. Suppose we want to numerically estimate the slope of the tangent line at $t = 8$ to the graph of the function that gives the number of polio cases in 1949: $y = \dfrac{42{,}183.911}{1 + 21{,}484.253 e^{-1.248911 t}}$

where $t = 1$ on January 31, 1949, $t = 2$ on February 28, 1949, and so forth.

Enter the polio cases equation in the y1 location of the y(x)= list. (Carefully check your entry of the equation, and be sure to use parentheses around the denominator and the exponent.) We now evaluate the slopes of secant lines that join close points to the left of $x = 8$ with $x = 8$.	Plot1 Plot2 Plot3 \y1■...3e^(-1.248911x))
On the home screen, type in the expression shown to the right to compute the slope of the secant line joining the close point where $x = 7.9$ and the point of tangency where $x = 8$.	(y1(8)-y1(7.9))/(8-7.9) 13159.6827248

Record on paper each slope, to at least 1 more decimal place than the desired accuracy, as it is computed. You are asked to find the nearest whole number that these slopes are approaching, so record at least one decimal place in your table of slopes.

Press 2nd ENTER (ENTRY) to recall the last entry, and then use the arrow keys to move the cursor over the 9 in the "7.9". Press 2nd DEL (INS) and press 9 to insert another 9 in <u>both</u> places that 7.9 appears. Press ENTER to find the slope of the secant line joining $x = 7.99$ and $x = 8$.	(y1(8)-y1(7.9))/(8-7.9) 13159.6827248 (y1(8)-y1(7.99))/(8-7.99) 13170.6176627
Continue in this manner, recording each result on paper, until you can determine to which value the slopes from the left seem to be getting closer and closer. It appears that the slopes of the secant lines from the left are approaching 13,170 cases per month.	13170.187131 (y1(8)-y1(7.9999))/(8-7.9999) 13170.12882 (y1(8)-y1(7.99999))/(8-7.99999) 13170.1229
We now evaluate the slopes joining the point of tangency and nearby close points to the *right* of $x = 8$. Clear the screen, recall the last expression with 2nd ENTER (ENTRY), and edit it so that the nearby point is $x = 8.1$. Press ENTER to calculate the secant line slope.	(y1(8)-y1(8.1))/(8-8.1) 13146.3842089

Continue in this manner as you did when calculating slopes to the left, but each time insert a 0 before the "1" in two places in the close point. Record each result on paper until you can determine the value the slopes from the right are approaching.

It appears that the slopes of the secant lines from the right are approaching 13,170 cases per month.

When the slopes from the left and the slopes from the right approach the same number, that number is the slope of the tangent line at the point of tangency. In this case, we estimate the slope of the tangent line to be 13,170 cases per month.

4.1.2 NUMERICALLY ESTIMATING SLOPES USING THE TABLE The process discussed in Section 4.1.1 can be done in fewer steps and with fewer keystrokes when you use the TI-86 TABLE. The point of tangency is $x = 8$, $y = $ y1(8), and let's call the close point $(x, y1(x))$. Then, slope $= \frac{\text{rise}}{\text{run}} = \frac{y1(x) - y1(8)}{x - 8}$. We illustrate numerically estimating the slope using the TABLE with the logistic function given in Section 4.1.1 of this *Guide*.

Have the polio cases equation given in the previous section in the y1 location of the y(x)= list. Enter the above slope formula in a different other location, say y2. (Remember to enclose both numerator and denominator of the slope formula in parentheses.) Turn off y1 because we are only considering the output from y2.

Press TABLE F2 [TBLST] and choose ASK in the Indpnt: location. (See page B-7 for more specific instructions.)

Access the table with F1 [TABLE] and delete or type over any previous entries in the x column. Enter values for x, the input of the close point, so that x gets closer and closer to 8 from the left.

- Although it did not happen in this illustration, the calculator might switch the input values you enter to scientific notation and display rounded output values so that the numbers can fit on the screen in the space allotted for outputs of the table. If this happens, you should position the cursor over each output value and record on paper as many decimal places as necessary in order to determine the limit to the desired degree of accuracy.

Repeat the process, entering values for x, the input of the close point, so that x gets closer and closer to 8 from the right.

Remember to highlight the output values and record on paper the slope to at least one more decimal position than the desired accuracy.

The limit is estimated to be the value that the limits from the left and the right are approaching: 13,170 cases per month.

NOTE: You may wish to leave the slope formula in y2 as long as you need it. Remember to change the input value at the point of tangency the next time you use this formula. Turn y2 off when you are not using it with F5 [SELCT].

4.2 Algebraically Finding Slopes

The TI-86 does not find algebraic formulas for slope, but you can use the built-in numerical derivatives and draw the graph of a derivative to check any formula that you find algebraically.

4.2.1 UNDERSTANDING YOUR CALCULATOR'S SLOPE FUNCTIONS
The TI-86 has two derivative functions: nDer (used in dxNDer mode) and der1 (used in dxDer1 mode). The numerical derivative nDer uses the slope of a secant line to approximate the slope of the tangent line at a point on the graph of a function. However, instead of using a secant line through the point of tangency and a close point, nDer uses the slope of a secant line through two close points that are equally spaced from the point of tangency.

The secant line joining the points $(a - k, f(a - k))$ and $(a + k, f(a + k))$ and the line tangent to the graph of f where $x = a$ are shown in Figure 4. Notice that the slopes of the secant line and tangent appear to be close to the same value even though these are different lines.

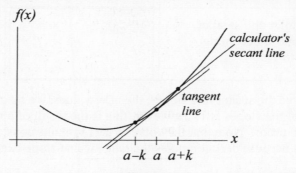

Figure 4

As k gets closer and closer to 0, the two points through which the secant line passes move closer and closer to a. Provided the slope of the tangent line exists, the limiting position of the secant line will be the tangent line.

The calculator's notation for the slope of the secant line shown in Figure 4 is

nDer(function, symbol for input variable, a)

The k that determines the locations of the points on the secant line can be set to different values, but specifying this value is optional. If it is not given, the calculator automatically uses $k = 0.001$ (or whatever the current setting may be). The TI-86 uses the symbol δ for k, and it also has a value called *tol*. This is the TI-86 tolerance – a quantity that can affect the speed and accuracy of some calculations.

The TI-86 also computes for many functions the slope of the tangent line. This derivative, der1, computes for many built-in functions the exact derivative. Even though der1 is more accurate than nDer, it is more restrictive because it does not work for all functions.

Any smooth, continuous function will do for our explorations, so let's investigate and compare nDer and der1 with the function in Example 3 of Section 4.2 of *Calculus Concepts*: $f(x) = 2\sqrt{x}$. First we find nDer, the slope of the secant line between the two points $(0, f(0))$ and $(2, f(2))$. That is, we are finding the slope of the line between the points $(a - k, f(a - k))$ and $(a + k, f(a + k))$ for $a = 1$ and $k = 1$. Enter $f(x) = 2\sqrt{x}$ in y1.

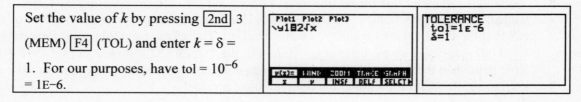

Set the value of k by pressing 2nd 3 (MEM) F4 (TOL) and enter $k = \delta = 1$. For our purposes, have tol = 10^{-6} = 1E–6.

On the home screen, access nDer(with 2nd ÷ (CALC) F2 [nDer], and enter nDer(y1, x, 1). Access der1(with F3 [der1] and enter der1(y1, x, 1).		nDer(y1,x,1)
		        1.41421356237
		der1(y1,x,1)
		                    1
		
		evalF nDer der1 der2 fnInt

Press 2nd 3 (MEM) F4 (TOL) and change the value of k from 1 to 0.1. On the home screen, recall the last two entries using 2nd ENTER (ENTRY).	TOLERANCE
	tol=1E-6
	δ=.1
		1.41421356237
		der1(y1,x,1)
		                    1
		nDer(y1,x,1)
		        1.0012555012
		der1(y1,x,1)
		                    1

Repeat the process for $k = 0.001$ and $k = 0.0001$. Note how the results are changing.	TOLERANCE
	tol=1E-6
	δ=.001
		1.0012555012
		der1(y1,x,1)
		                    1
		nDer(y1,x,1)
		        1.00000012505
		der1(y1,x,1)
		                    1

As k becomes smaller and smaller, the secant line slope is becoming closer and closer to 1.	TOLERANCE
	tol=1E-6
	δ=.0001
		1.00000012505
		der1(y1,x,1)
		                    1
		nDer(y1,x,1)
		        1.0000000015
		der1(y1,x,1)
		                    1

- When we observe the secant line slopes, the nDer values, we reach the conclusion that the limiting value of the slopes is 1. However, this is the result of a numerical investigation, not an algebraic proof. Note that the value of der1 remains constant at 1 no matter what k is used. This is because der1 computes the tangent line slope exactly for this function.

In the table below, the first row lists some values of a, the input of a point of tangency, and the second row gives the actual slope (to 7 decimal places) of the tangent line at those values. The algebraic method gives the exact slope of the line tangent to the graph of f at these input values.

Use your calculator to verify the values in the third through sixth rows that give the values of der1 and nDer (to 7 decimal places), the slope of the secant line between $(a - k, f(a - k))$ and $(a + k, f(a + k))$ for the indicated values of k. Before calculating each secant line slope, set the value of $k = \delta$ on the TOLERANCE screen accessed with 2nd 3 (MEM) F4 (TOL).

a = input of point of tangency	2.3	5	12.82	62.7
slope of tangent line = $f'(a)$	0.6593805	0.4472136	0.2792904	0.1262892
value of der1(y1, x, a)	0.6593805	0.4472136	0.2792904	0.1262892
slope of secant line, $k = 0.1$	0.6595364	0.4472360	0.2792925	0.1262892
slope of secant line, $k = 0.01$	0.6593820	0.4472138	0.2792904	0.1262892
slope of secant line, $k = 0.001$	0.6593805	0.4472136	0.2792904	0.1262892
slope of secant line, $k = 0.0001$	0.6593805	0.4472136	0.2792904	0.1262892

You can see that der1(y1, x, a) gives the slope of the tangent line and that the secant line slope is very close to the slope of the tangent line for small values of k. For this function, the secant line slope does a great job of approximating the slope of the tangent line when k is very small.

Will the slope of this secant line always do a good job of approximating the slope of the tangent line when $k = 0.001$? Yes, it generally does, as long as the instantaneous rate of change exists at the input value at which you evaluate nDer. Therefore, from this point forward we use $k = 0.001$ whenever we evaluate nDer. Return to the TOLERANCE screen and set $k = \delta = 0.001$. Note that when the instantaneous rate of change does not exist, neither nDer nor der1 should be used to approximate something that does not have a value!

You will benefit from reading again Section 3.2.4 of this *Guide*, which illustrates several cases when the TI-86's slope function nDer gives a value for the slope when it does not exist. For instance, consider $f(x) = |x|$ for which the instantaneous rate of change does not exist at $x = 0$ because the graph of f has a sharp point there. Note that you can enter the function in the y(x)= list or enter it directly into the derivative instruction on the home screen.

Find the TI-86's derivatives at $x = 0$ for $f(x) = |x|$. Note that the slope of the secant line is 0, but the instantaneous rate of change does not exist at $x = 0$.	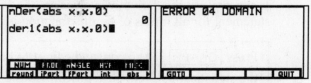

- When the der1 instruction is entered, it gives an error message (as it should). Whenever you receive an error message, you need to carefully examine what you are doing and see if there is a mathematical reason for the error message (as is the case here).

We recommend that you keep the TI-86's MODE to those settings given in Figure 1 on page B-1. Check that your calculator has those settings before proceeding. Because it is more mathematically accurate, we also suggest that you use der1 rather than nDer whenever possible.

4.4 More Simple Rate-of-Change Formulas

The TI-86 only approximates or calculates numerical values of slopes – it does not give the slope in formula form. You also need to remember that nDer computes slope by a different definition than we use and that you may not obtain the correct answer when using nDer with functions at points where the slope does not exist. For this reason we use der1, not nDer.

4.4.1 DERIVATIVE NOTATION AND CALCULATOR NOTATION In addition to learning when your calculator gives an acceptable answer for a derivative and when it does not, you also need to understand the differences and similarities in mathematical derivative notation and calculator notation. The notation that we use for the calculator's numerical derivative is der1($f(x), x, x$). The correspondence between the mathematical derivative notation $\frac{df(x)}{dx}$ and the calculator's notation der1($f(x), x, x$) is illustrated in Figure 5.

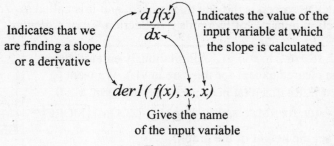

Figure 5

We illustrate another use of the TI-86's derivative by constructing Table 4.14 in Section 4.4 of *Calculus Concepts*. The table lists x, $y = f(x) = e^x$, and $y' = \frac{df}{dx}$ for seven different inputs.

You can evaluate the TI-86 derivative on the home screen or in the table. We choose to use the table. Enter $f(x) = e^x$ in the y(x)= list, say in y1. In y2 enter the numerical derivative evaluated at a general input x.	

| Press [TABLE] [F2] [TBLST], and choose ASK in the Indpnt: location. Access the table with [F1] [TABLE] and enter -3, -2, -1, 0, 1, 2, and 3 for x. |

It appears that the derivative values are the same as the function outputs. In fact, this is a true statement for all inputs of f – this function is its own derivative!

| Press [GRAPH] [F1] [y(x)=] and edit y1 to be $g(x) = 2^x$. Access the statistical lists and clear any previous entries from L1, L2, L3, and L4. Enter the x-values shown above in L1. Highlight L2 and type y1(L1). |

| Press [ENTER] to fill L2 with the function outputs. Then, highlight L3 and type y2(L1). |

| Press [ENTER] to place in list L3 the derivative of y1 evaluated at the inputs in L1. Note that these values are not the same as the function outputs. To see what relation the slopes have to the function outputs, press [▶] and highlight L4. Enter L3 ÷ L2. |

It appears that the slope values are a multiple of the function output. In fact, that multiple is $\ln 2 \approx 0.693147$. The slope formula is: If $g(x) = 2^x$, then $\frac{dg}{dx} = (\ln 2) 2^x$.

4.4.2 CALCULATING $\frac{dy}{dx}$ AT SPECIFIC INPUT VALUES

The previous two sections of this *Guide* examined the calculator's numerical derivatives, nDer($f(x)$, x, a) and der1($f(x)$, x, a), and illustrated that they equal or give a good approximation to the slope of the tangent line at points where the instantaneous rate of change exists. You can also evaluate the calculator's derivatives from the graphics screen using the CALC menu. Whichever derivative you have selected in the MODE screen is the one that is used, and it is called dy/dx. We illustrate this use with the function in part a of Example 2 in Section 4.4 of *Calculus Concepts*.

| Enter $f(x) = 12.36 + 6.2 \ln x$ in y1. (All previously entered functions should be cleared except for this one in y1.) We want to draw a graph of f. Realize that because of the log term, $x \geq 0$. Choose some value for xMax, say 5. Use [F3] [ZOOM] [MORE] [F1] [ZFIT] to set the height of the graph. |

| With the graph on the screen, press [MORE] [F1] [MATH] [F2] [dy/dx]. Use [◀] or [▶] to move to some point on the graph. |

| Press [ENTER] and the slope of the function is calculated at the input you chose in the previous step. |

TI-86 Guide   B-57

If you want the derivative evaluated at a specific value of x, you can type in the desired input instead of pressing the arrow keys. Press [CLEAR] [EXIT] (to return the menu) and [F1] [MATH] [F2] [dy/dx] 3 [ENTER].
The slope $dy/dx \approx 2.066667$ appears at the bottom of the screen. Return to the home screen and press [x-VAR]. The TI-86 x memory location has been updated to 3. Now type both numerical derivative instructions (evaluated at 3) as shown to the right. This dy/dx value you saw on the graphics screen is the der1 value because we have the MODE set to dxDer1.

You can use the ideas presented above to check your algebraic formula for the derivative. We next investigate this procedure.

4.4.3 NUMERICALLY CHECKING SLOPE FORMULAS

It is always a good idea to check your answer. Although your calculator cannot give you an algebraic formula for the derivative function, you can use numerical techniques to check your algebraic derivative formula. The basic idea of the checking process is that if you evaluate your derivative and the TI-86's derivative at several randomly chosen values of the input variable and the outputs are basically the same values, your derivative is *probably* correct.

These same procedures are applicable when you check your results (in the next several sections) after applying the Sum Rule, the Chain Rule, or the Product Rule. We use the function in part *c* of Example 2 in Section 4.4 of *Calculus Concepts* to illustrate.

Enter $m(r) = \frac{8}{r} - 12\sqrt{r}$ in y1 (using x as the input variable).
Compute $m'(r)$ using pencil and paper and the derivative rules. Enter this function in y2. (What you enter in y2 may or may not be the same as what appears to the right.)
Enter the TI-86's derivative of y1 (evaluated at a general input x) in y3. Because you are interested in seeing if the outputs of y2 and y3 are the same, turn off y1.
Press [TABLE] [F2] [TBLST] and choose ASK in the Indpnt: location. Access the table with [F1] [TABLE] and enter at least three different values for x.

The table gives fairly strong evidence that that y2 and y3 are the same function.

4.4.4 GRAPHICALLY CHECKING SLOPE FORMULAS

When it is used correctly, a graphical check of your algebraic formula works well because you can look at many more inputs when drawing a graph than when viewing specific inputs in a table. We illustrate this use with the function in part *d* of Example 2 in Section 4.4 of *Calculus Concepts*.

Enter $j(y) = 17{,}000\left(1 + \frac{0.025}{12}\right)^{12y}$ in y1 (using x as the input variable). Enter your answer for $j'(y)$ in y2 and the TI-86 derivative of y1 (evaluated at a general input x) in y3.

NOTE: y1 = 17000(1 + 0.025/12)^(12x) and we entered y2 = 17000*ln(1.025288)(1.025288^x).

Before proceeding, turn off the graphs of y1 and y3. To graphically check your derivative formula, you now need to find a good graph of y2. Because this function is not in a context with a given input, the time it takes to draw a graph is shortened if you know the approximate shape of the graph. Note that the graph of the function in y2 is an increasing exponential curve.

Start with F3 [ZOOM] MORE F4 [ZDECM] or F3 [ZOOM] F4 [ZSTD]. Neither of these shows a graph (because of the large coefficient in j), but you can press TRACE to see some of the output values. Use those values to reset the window.

- The graph shown above was drawn in the window [−10, 10] by [330, 550], but any view that shows a graph will do.

Now, press F1 [y(x)=], turn off y2, and turn on y3, which is the derivative computed by the TI-86. Press F5 [GRAPH]. Note that you are drawing the graph of y3 in *exactly the same window* in which you graphed y2.

If you see the same graph, your algebraic formula (in y2) is very likely correct.

4.5 The Chain Rule

You probably noticed that checking your answer for a slope formula graphically is more difficult than checking your answer using the TI-86 table if you have to spend a lot of time finding a window in which to view the graph. Practice with these methods will help you determine which is the best to use to check your answer. (These ideas also apply to Section 4.6.)

4.5.1 SUMMARY OF CHECKING METHODS Before you begin checking your answer, make sure that you have correctly entered the function. It is very frustrating to miss the answer to a problem because you have made an error in entering a function in your calculator. We summarize the methods of checking your algebraic answer using the function in Example 3 of Section 4.5 of *Calculus Concepts*.

Enter the function $P(t) = \dfrac{77.9}{1+39.5e^{-0.4984t}}$ in the y1 location, your answer for P' in y2, and the TI-86 derivative in y3. Turn off y1. Note that your entry in y2 may be different, but we entered y2 = 1533.602e^(−0.4984x)/(1+39.5e^(−0.4984x))^2.	
Go to the table, which has been set to ASK mode. Enter at least 3 input values. (Because this problem is in a context, read the problem to see which inputs make sense.) It seems that the answer in y2 is probably correct!	
If you prefer a graphical check, the problem states that the equation is valid between 1980 and 1996 (and the input is the number of years after 1980). So, turn off y3, set xMin = 0, xMax = 16, and draw a graph of y2 using ZFit. Then, using the same window, draw a graph of y3 with y2 turned off. The graphs are the same, again indicating a correct answer in y2.	

Chapter 5   Analyzing Change: Applications of Derivatives

5.2 Relative and Absolute Extreme Points

Your calculator can be very helpful for checking your analytic work when you find optimal points and points of inflection. When you are not required to show work using derivative formulas or when an approximation to the exact answer is all that is required, it is a simple process to use your calculator to find optimal points and inflection points.

5.2.1 FINDING X-INTERCEPTS OF SLOPE GRAPHS
Where the graph of a function has a relative maximum or minimum, the slope graph has a horizontal tangent. Where the tangent line is horizontal, the derivative of the function is zero. Thus, finding where the slope graph *crosses* the input axis is the same as finding the input of a relative extreme point.

Consider, for example, the model for Acme Cable Company's revenue for the 26 weeks after it began a sales campaign, where x is the number of weeks since Acme began sales:

$$R(x) = -3x^4 + 160x^3 - 3000x^2 + 24{,}000x \text{ dollars}$$

In Example 2 of Section 5.2 of *Calculus Concepts*, we are asked to find when Acme's revenue was greatest during the 26-week interval.

Enter R in the y1 location of the y(x)= list. Enter either the TI-86's derivative or your derivative in the y2 location. Turn off y1. (If you use your derivative, be sure to use one of the methods at the end of Chapter 4 in this *Guide* to check that your derivative and the TI-86 derivative are the same.)	
The statement of the problem indicates that x should be graphed between 1 and 26. Set this horizontal view, and draw the slope graph in y2 with F3 [ZOOM] MORE F1 [ZFIT]. Reset the window to a better view for this illustration: yMin = -800 and yMax = 3000. Redraw the slope graph with F5 [GRAPH].	
With the graph on the screen, find the intercepts of the slope graph with MORE F1 [MATH] F1 [ROOT]. Press and hold ▶ to move the cursor near to, but still to the *left* of, the rightmost x-intercept.	
Press ENTER to mark the location of the left bound. Use ▶ to move the cursor near to, but to the *right* of, the rightmost x-intercept.	
Press ENTER to mark the location of the right bound. You are next asked to provide a guess. Any value near the intercept will do. Use ◀ to move the cursor near the intercept.	

- Note that the calculator has marked (at the top of the screen) the left and right bounds with small triangles. The x-intercept must lie between these two bound marks. If you incorrectly mark the interval, you may not get an answer.

Press ENTER. The location of the x-intercept is displayed. We see that $R'(x) = 0$ at $x = 20$. (The numerical process used the find the root sometimes includes small rounding errors. The output (y) at the root should be 0 but is printed on the screen as -0.000000005.)	

We now need to determine if the slope graph crosses the x-axis, only touches the x-axis, or does neither of these at the other location that may be an intercept.

Zoom in with ZIN (see Section 1.4.2) or BOX (see Section 3.2.1) as many times as necessary to magnify the portion of the graph around $x = 20$ in order to examine it more closely. We choose to use BOX, but both work equally well.	
After magnifying the graph several times, we see that the graph just touches and does not cross the x-axis near $x = 10$. (Press F4 [TRACE] and trace as near as possible to $x = 10$.) We see that $x = 10$ does not yield an extreme point on the graph of R.	
We are asked to find the absolute maximum, and we know that it occurs at one of the endpoints of the interval or at a zero of the slope graph. So, return to the home screen and find the outputs of R at the endpoints of the interval ($x = 0$ and $x = 26$) as well as the output at the "crossing" root of the slope graph ($x = 20$).	

We see that Acme's revenue was greatest at 20 weeks after they began the sales campaign.

5.2.2 FINDING ZEROS OF SLOPE FUNCTIONS USING THE SOLVER

You may find it more convenient to use the TI-86 SOLVER rather than the graph and the CALC menu to find the x-intercept(s) of the slope graph. We illustrate using Acme Cable's revenue function.

Enter R in the y1 location of the y(x)= list. Enter either the TI-86's derivative or your derivative in the y2 location. Turn off y2.	
Draw a graph of y1 to obtain a guess as to the location of any relative maxima for use in the SOLVER. We are told that x is between 1 and 26. Set this horizontal view, and draw the graph of y1 with F3 [ZOOM] MORE F1 [ZFIT].	
Reset yMax to a larger value, say 95,000, to better see the high point on the graph. Graph R and press TRACE. Hold down ▶ until you have an estimate of the input location of the high point. The maximum seems to occur when x is near 19.8.	

Remember that we want to solve the equation $R'(x) = 0$ to find where the tangent is horizontal. We are using as a guess of the answer the value we obtained by tracing the graph of R.

Access the SOLVER with [2nd] [GRAPH] (SOLVER). Clear what is there from a previous problem, type y2 as the left side of the equation, and press [ENTER]. Enter 0 as the right-hand side of the equation; type the guess $x \approx 19.8$. With the cursor on the line containing x, press [F5] [SOLVE]. The solution $x = 20$ is found.

NOTE: Recall that calculators use numerical algorithms to find zeros. You may or may not obtain the exact value 20. Always round the answer obtained from the SOLVER so that it makes sense in the problem context (here, round to a whole number.)

Note that if you enter $x \approx 10$ as a guess in the SOLVER, it returns the solution $x = 10$. You should not rely only on the SOLVER – you must also examine the function graph to see that only 20 is the location of the maximum. Acme's revenue was greatest at 20 weeks after they began the sales campaign.

5.2.3 USING THE CALCULATOR TO FIND OPTIMAL POINTS

Once you draw a graph of a function that clearly shows the optimal points, your calculator can find the location of those high points and low points without using calculus. However, we recommend not relying only on this method because your instructor may ask you to show your work using derivatives. If so, this method would probably earn you no credit! This method does give a good check of your answer, and we illustrate it using Acme Cable's revenue function R from Section 5.2.1.

Enter R in the y1 location of the y(x)= list. The problem indicates that x should be graphed between 1 and 26, so set this horizontal view. Draw the function graph with [F3] [ZOOM] [MORE] [F1] [ZFIT]. Press [F2] [WIND] and reset yMax to 95,000. Press [F5] [GRAPH].

With the graph of R on the screen, press [MORE] [F1] [MATH] [F5] [FMAX]. Use [▶] to move the cursor near, but still to the *left* of, the high point on the curve.

Press [ENTER] to mark the left bound of the interval. Use [▶] to move the cursor to the *right* of the high point on the curve.

Press [ENTER] to mark the right bound of the interval. Use [◀] to move the cursor to your guess for the high point on the curve.

CAUTION: Notice the small arrowheads at the top of the screen that mark the bounds of the interval. The TI-86 returns the highest point that is within the bounded interval that you have marked when you press ENTER . If the relative maximum does not lie in this interval, the TI-86 will return, instead, the highest point in the interval you marked (usually an endpoint).

Press ENTER and the location of the relative maximum (x) and the relative maximum value (y) are displayed. (As previously mentioned, you may not see the exact answer due to rounding errors in the numerical routine used by the TI-86. Always round your answer to make sense in the context.)	

- The method shown in this section also applies to finding the relative minimum values of a function. The only difference is that to find the relative minimum instead of the relative maximum, initially press GRAPH MORE F1 [MATH] F4 [FMIN].

5.4 Inflection Points

As was the case with optimal points, your calculator can be very helpful in checking algebraic work when you find points of inflection. You can also use the methods illustrated in Section 5.2.3 of this *Guide* to find the location of any maximum or minimum points on the graph of the first derivative to find the location of any inflection points for the function.

5.4.1 FINDING X-INTERCEPTS OF A SECOND-DERIVATIVE GRAPH We first look at using the algebraic method of finding inflection points – finding where the graph of the second derivative of a function *crosses* the input axis.

To illustrate, we consider a model for the percentage of students graduating from high school in South Carolina from 1982 through 1990 who entered post-secondary institutions:

$$f(x) = -0.1057x^3 + 1.355x^2 - 3.672x + 50.792 \text{ percent}$$

where x is the number of years after 1982.

Enter f in the y1 location of the y(x)= list, the first derivative of f in y2, and the second derivative of f in y3. (Be careful not to round any decimal values.) Turn off y1 and y2.	
We are given the input interval 1982 through 1990, so $0 \leq x \leq 8$. Either use ZFIT or choose some appropriate vertical view. We use $-4 \leq y \leq 4$. Because we are looking for the x-intercept(s) of the second derivative graph, any view that shows the line crossing the horizontal axis is okay to use.	
Use the methods indicated in Section 5.2.1 to find where the second derivative graph crosses the x-axis. (Note that when you are asked to give the inflection *point* of f, you should answer with both the input and an output of the original function.)	
Return to the home screen and enter x-VAR . The x-value you just found as the x-intercept remains stored in the x location until you change it by tracing, using the SOLVER, and so forth. Find the y-value by substituting this x-value into y1.	

TI-86 Guide   B-63

> At some point, be sure to examine a graph of the function and verify that an inflection point does occur at the point you have found. To do this, turn off y3, turn on y1, and use ZFIT to draw the graph. Trace near where $x \approx 4.27$ and $y \approx 51.6$. The graph of R confirms that an inflection point occurs at this point.

EXPLORE: What would the line tangent to the graph of f look like at the inflection point? Use the graph of y1 and the first method explained on page B-45 of this *Guide* (with $x = 4.27$) to see if you are correct.

EXPLORE: What is true about the graph of f′ at the inflection point? Use the graph of y2 and the trace cursor to see if you are correct.

- The TI-86 usually draws an accurate graph of the first derivative of a function when you use der1. It also draws the graph of and computes numerically the second derivative of many functions with a command called der2. If you are in doubt about your calculation of the first and second derivative formulas, use the calculator's derivatives to avoid making a mistake in the location of the inflection point.

> Enter f in the y1 location of the y(x)= list, the first derivative of f in y2, and the second derivative of f in y3, using the TI-86's derivative for each derivative that you enter. Access der2 with [2nd] [÷] (CALC) [F4] [der2]. Turn off y1. Be sure that you set xMin = 0 and xMax = 8 (as the problem directions indicate.)
>
> Draw the graph of the first derivative (y2) and the second derivative (y3) of f using an appropriate window. You can use ZFIT to set the vertical view or experiment until you find a suitable view. The graph to the right uses $-4 \le y \le 4$.

Find the x-intercept of the second derivative graph as indicated in this section or find the input of the high point on the first derivative graph (see Section 5.2.3) to locate the input of the inflection point. (When more than one graph is on the screen, use [▼] to jump between graphs.)

5.4.2 USING THE CALCULATOR TO FIND INFLECTION POINTS

Remember that an inflection point on the graph of a function is a point of greatest or least slope. Whenever finding the second derivative of a function is tedious algebraically and/or you do not need an exact answer from an algebraic solution, you can easily find the input location of an inflection point by finding where the first derivative of the function has a maximum or minimum slope.

We illustrate this method using the logistic function for polio cases that is in Example 2 of Section 5.4 in *Calculus Concepts*:

The number of polio cases in the U.S. in 1949 is given by $C(t) = \dfrac{42{,}183.911}{1 + 21{,}484.253 e^{-1.248911 t}}$

where $t = 1$ in January, $t = 2$ in February, and so forth.

> Enter $C(x) = 42183.911/(1 + 21484.253 e^{\wedge}(-1.248911x))$ in the y1 location of the y(x)= list and the first derivative of f in y2. (You can use your algebraic formula for the first derivative or the calculator's derivative.) Turn off y1.

The problem context says that the input interval is from 0 (the beginning of 1949) to 12 (the end of 1949), so set these values for x in the window. Set the vertical view and draw the graph of C' with ZFIT.	
Use the method discussed in Section 5.2.3, *i.e.*, MORE F1 [MATH] F5 [FMAX], to find the input location of the maximum point on the slope graph.	
The x-value of the maximum of the slope graph is the x-value of the inflection point of the function. To find the rate of change of polio cases at this time, substitute this value of x in y2. To find the number of cases at this time, substitute x in y1.	

CAUTION: Do not forget to round your answers appropriately (this function should be interpreted discretely) and to give units of measure with each answer.

Note that you could have found the input of the inflection point on the polio cases graph by finding the x-intercept of the second derivative graph. The function whose graph is shown to the right is y3 = der2(y1, x, x), and the graph was drawn using ZFIT.	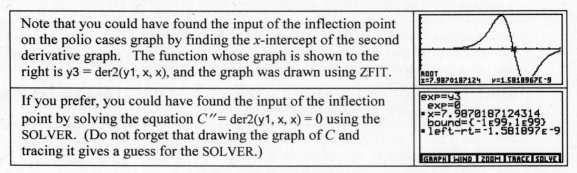
If you prefer, you could have found the input of the inflection point by solving the equation $C'' =$ der2(y1, x, x) = 0 using the SOLVER. (Do not forget that drawing the graph of C and tracing it gives a guess for the SOLVER.)	

Our final method is also the simplest – using the TI-86's built-in inflection point finder! Be certain if you use this method that the function does have an inflection point at the location indicated by the calculator. We illustrate with the polio cases function that is entered in y1: $C(x) = 42183.911/(1 + 21484.253e^{\wedge}(-1.248911x))$.

Turn y1 on and draw a graph of C between $x = 0$ and $x = 12$ with ZFIT. Press MORE F1 [MATH] MORE F1 [INFLC]. Move, using ▶ or ◀, the cursor to a position to the left of where the function changes concavity.	
Press ENTER to mark the left bound. Use ▶ to move the cursor to the right of where the function changes concavity. Press ENTER to mark the right bound. At the Guess? Prompt, move the cursor to the approximate location of the inflection point.	
Press ENTER and the input and output of the inflection point are displayed.	

Chapter 6 Accumulating Change: Limits of Sums and the Definite Integral

6.1 Results of Change and Area Approximations

So far, we have used the TI-86 to investigate rates of change. In this chapter we consider the second main topic in calculus – the accumulation of change. You calculator has many useful features that will assist in your investigations of the results of change.

6.1.1 AREA APPROXIMATIONS USING LEFT RECTANGLES

The TI-86 lists can be used to approximate, using left rectangles, the area between the horizontal axis and a rate of change function between two specified input values. We illustrate the necessary steps using the data for the number of customers entering a large department store during a Saturday sale. These data appear in Table 6.1 of Section 6.1 in *Calculus Concepts*.

Minutes after 9 a.m.	0	45	75	120	165	195	255	330	370	420	495	570	630	675
Customers per minute	1	2	3	4	4	5	5	5	5	4	4	3	2	2

Enter these data in lists L1 and L2. Find a cubic function to fit the data, and paste the function in y1. If you did not obtain this function (shown below rounded to 4 decimal places), check your data entry: $c(x) = (4.5890 \cdot 10^{-8})x^3 - (7.7813 \cdot 10^{-5})x^2 + 0.0330x + 0.8876$ First, we graph the function on a scatter plot of the data over the 12-hour (720 minute) sale.

Delete any functions in the y(x)= list and turn on Plot 1. Press [F2] [WIND], either set xMin = 0, xMax = 720, xScl = 60 and use ZFIT to set the vertical view or use [F3] [ZOOM] [MORE] [F5] [ZDATA] to draw a graph. Set yMin = 0 and press [F5] [GRAPH].	

We want to approximate the area under the cubic graph and above the *x*-axis during the 12-hour time period using rectangles of equal width (60 minutes).

Press [2nd] [+] (STAT) [F2] [EDIT], clear L1 and L2, and enter the inputs 0, 60, 120, …, 720 in L1. A quick way to do this is to highlight L1 and press [2nd] [−] (LIST) [F5] [OPS] [MORE] [F3] [seq(] [x-VAR] [,] [x-VAR] [,] 0 [,] 720 [,] 60 [)]. (Section 2.1.1 gives a more detailed explanation of the sequence command.)	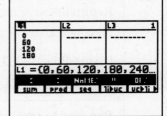
Because we are using rectangles with heights determined at the left endpoint of each interval, delete the last value in L1 (720). Substitute each input in L1 into the function in y1 by highlighting L2 and typing y1(L1). Press [ENTER].	

Consider what is now in the lists. L1 contains the left endpoints of the 12 rectangles and L2 contains the heights of the rectangles. If we multiply the heights by the widths of the rectangles (60 minutes) and enter this product in L3, we will have the rectangle areas in L3.

Copyright © Houghton Mifflin Company. All rights reserved.

Highlight L3 and type 60 [X] [ALPHA] 7 (L) 2 [ENTER]. L3 now contains the areas of the 12 rectangles.

All that remains is to add the areas of the 12 rectangles. To do this, return to the home screen and type [2nd] [–] (LIST) [F5] [OPS] [MORE] [F1] [sum] [ALPHA] 7 (L) 3 [ENTER].

We estimate that 2574 customers came to the Saturday sale.

6.1.2 AREA APPROXIMATIONS USING RIGHT RECTANGLES

When you use left rectangles to approximate the results of change, the rightmost data point is not the height of a rectangle and is not used in the computation of the left-rectangle area. Similarly, when using right rectangles to approximate the results of change, the leftmost data point is not the height of a rectangle and is not used in the computation of the right-rectangle area. We illustrate the right-rectangle approximation with the data in Table 6.3 in Example 2 of Section 6.1.

End of day	1	5	9	13	17	21	25	29
Concentration rate of change (μg/mL/day)	1.5	0.75	0.33	0.20	0.10	−1.1	−0.60	−0.15

The data show the rate of change of the concentration of a drug in the blood stream in terms of the number of days since the drug was administered. Part *a* of Example 2 says to use the data and right rectangles of width 4 days to estimate the change in concentration of the drug from the end of day 1 through the end of day 29. Enter these data in lists L1 and L2.

Because the leftmost data point is not used with a right rectangle approximation, delete (1, 1.5) from lists L1 and L2. List L2 now contains the (signed) heights of the right rectangles. Highlight L3 and press 4 [X] [ALPHA] 7 (L) 2 [ENTER]. List L3 now contains the (signed) rectangle areas.

Return to the home screen and find the sum of the seven right-rectangle areas with type [2nd] [–] (LIST) [F5] [OPS] [MORE] [F1] [sum] [ALPHA] 7 (L) 3 [ENTER].

The change in concentration was about −1.88 μg/mL.

Part *b* of Example 2 says to find a model for the data. Return to the lists. Insert the first data point (1, 1.5) by placing the cursor over the first entries in L1 and L2 and pressing [2nd] [DEL] (INS). Type the data value over the 0 that is inserted in each.

A scatter plot shows that a piecewise model with a break point at $x = 20$ is appropriate.

Delete the last three data points from lists L1 and L2.

TI-86 Guide   B-67

Fit an exponential model to the data for days 1 through 17 (the five remaining data values in L1, L2), and put the function in y1.	ExpReg y=a*b^x a=1.70819479 b=.844976548 ↓corr=-.9977036
Type in the data for days 21, 25, and 29 in lists L3 and L4. Find a linear function for these data and paste the equation in y2 with [2nd] [+] (STAT) [F1] [CALC] [F3] [LinR] [,] [ALPHA] 7 (L) 3 [,] [ALPHA] 7 (L) 4 [,] [2nd] [ALPHA] (alpha) 0 (y) 2.	LinReg y=a+bx a=-3.5854167 b=.11875 ↓corr=.999538639

Refer to Section 1.3.8 of this *Guide* if you want to review the instructions for graphing this piecewise continuous function on a scatter plot of the data.

Part *c* of Example 2 says to find the change in the drug concentration from $x = 0$ through $x = 20$ using right rectangles of width 2 days. Enter the right endpoints (2, 4, 6, ..., 20) in L1. Use the exponential function in y1 to find the rectangle heights. Enter the heights in L2.	L1: 2,4,6,8,10,12 L2: 1.219626, .8707952, .621735, .4438097, .316945, .2262941 L2 =y1(L1)
Find the right-rectangle areas by multiplying each entry in L2 by 2. This is the same as multiplying the sum of the heights by 2. On the home screen, press 2 [×] [2nd] [–] (LIST) [F5] [OPS] [MORE] [F1] [sum] [ALPHA] 7 (L) 2 [ENTER].	2*sum L2 8.23481364815

6.1.3 AREA APPROXIMATIONS USING MIDPOINT RECTANGLES
Areas of midpoint rectangles are found using the same procedures as those used to find left and right rectangle areas except that the midpoint of the base of each rectangle is in L1 and no data values are deleted. We illustrate the midpoint-rectangle approximation using the data in Table 6.6 in Example 3 of Section 6.1. First, we align the input of Table 6.6 as shown below.

Number of hours since 11:45 a.m. Wednesday	0	4	8	12	16	20
Flow rate (cfs)	785	753	834	1070	1470	2090

Enter the data in the above table into lists L1 and L2. Because area units are height units times width units, we need to convert the output data in L2 to cubic feet per hour by multiplying each value by 3600 (the number of seconds in one hour).	L1: 0,4,8,12,16,20 L2: 785,753,834,1070,1470,2090 L2 =3600*L2
Press [ENTER]. After conversion, the units for the area are $\frac{\text{cubic feet}}{\text{second}} \cdot \frac{3600 \text{ seconds}}{1 \text{ hour}} \cdot$ hours $\Rightarrow$ cubic feet. A scatter plot of the converted data shows that a quadratic function is appropriate.	(scatter plot)
Find the quadratic equation and paste it into y1. To use 10 intervals and midpoint rectangles to approximate the area under the graph of the equation from $x = 0$ through $x = 20$, the width of each rectangle is 2 hours.	QuadraticReg y=ax²+bx+c n=6 PRegC= {18225 -135334.28571...

Enter the midpoints 1, 3, 5, ..., 19 in L3. The heights of the rectangles are determined by the function in y1, so evaluate y1 at each of the values in L3 to find the rectangle heights.

You can multiply each rectangle height by the width (2 hours), enter these values in another list, and sum the rectangle areas or you can multiply the sum of the heights in L4 by 2. Either calculation results in the area estimate. (To access the sum instruction, press [2nd] [–] (LIST) [F5] [OPS] [MORE] [F1] [sum].)

6.1.4 SIMPLIFYING AREA APPROXIMATIONS

The procedures used in the previous three sections of this *Guide* can become tedious when the number of rectangles is large. When you have a function $y = f(x)$ in the y1 location of the y(x)= list, you will find program NUMINTGL very helpful in determining left-rectangle, right-rectangle, and midpoint-rectangle approximations for accumulated change. Program NUMINTGL (listed in the *TI-86 Program Appendix*) performs automatically all the calculations that you have been doing manually using the lists.

WARNING: This program will not execute properly unless the function that determines the heights uses x as the input variable and is pasted in y1. If you receive an error message while attempting to run this program, consult Programs in *TroubleShooting the TI-86* in this *Guide*.

We illustrate using this program with the function for the Carson River flow rates that was fit to data in the previous section.

$$f(h) = 18{,}225h^2 - 135{,}334.3h + 2{,}881{,}542.9 \text{ cubic feet per hour}$$

h hours after 11:45 a.m. Wednesday. (The complete function found from the data in Table 6.6 is used for all following calculations. It should be in the y1 location of the y(x)= list.)

Press [PRGM] [F1] [NAMES] and the F-key under NUMIN. Press [ENTER] to start program NUMINTGL. If you did not enter the function in y1, press [F2] [No] to exit the program. Enter the function in y1 and re-run the program. If the function is in y1, press [F1] [Yes] to continue.

At the next prompt, press [F1] [Yes] to draw the approximating rectangles or press [F2] [No] to obtain only numerical approximations to the area between the function and the horizontal axis between 0 and 20. For illustration purposes, let's choose to see the rectangles.

At the Left endpoint? prompt, press 0 [ENTER], and at the right endpoint? prompt, press 20 [ENTER] to tell the TI-86 the input interval. You are next shown a menu of choices. Press 4 and [ENTER] for a midpoint-rectangle approximation.

Input 10 at the N? prompt and press ENTER. A graph of the 10 midpoint rectangles and the function are shown. (Note that the program automatically sets the height of the window based on the left and right endpoints of the input interval.)	
The midpoint-rectangle area is displayed as SUM when you press ENTER.	Right Rect (2) Trapezoids (3) Midpt Rect (4) 4 N? 10 SUM= .79042500
Press ENTER once more and some choices are displayed. Suppose you now want to find the approximating area and see the figure for 10 right rectangles. Press 2 to change the method, and then press 2 ENTER to choose right rectangles.	Enter Choice: Change N (1) Change Method (2) Quit (3) 2
Press 10 ENTER at the N? prompt and view the 10 right rectangles. Again press ENTER and the right-rectangle area estimate is displayed.	Right Rect (2) Trapezoids (3) Midpt Rect (4) 2 N? 10 SUM= .83990314.2857
Continue in this manner to find the left-rectangle area or change N and find the left-, right-, or midpoint-rectangle approximations for different numbers of subintervals. When you finish, press ENTER and choose 3 to exit the program.	Enter Choice: Change N (1) Change Method (2) Quit (3) 3

   ## 6.2 The Definite Integral as a Limit of Sums

This section introduces you to a very important and useful concept of calculus – the definite integral. Your calculator can be very helpful as you study definite integrals and how they relate to the accumulation of change.

6.2.1 LIMIT OF SUMS

When you are looking for a trend in midpoint-rectangle approximations to the area (or signed area) between a function and the horizontal axis between two values of the input variable, program NUMINTGL is extremely useful! However, when finding a limit of sums using this program, it is not advisable to draw pictures when N, the number of rectangles, is large unless you have a lot of time to wait for many, many rectangles to be drawn.

We illustrate using this program to find a limit of sums using $f(x) = \sqrt{1-x^2}$, the function in Example 1 of Section 6.2 of *Calculus Concepts*.

To construct a table of midpoint-rectangle approximations for the area between f and the x-axis from $x = 0$ to $x = 1$, first clear the $y(x)=$ list, enter f in y1, and turn off Plot 1. (Type the square root symbol with 2nd x^2.)	Plot1 Plot2 Plot3 \y18√(1-x²)
Run program NUMINTGL. At the first prompt, press F1 [Yes] to continue and F2 [No] to not draw pictures. At the Left endpoint? prompt, enter 0, and at the Right endpoint? prompt, enter 1 to tell the TI-86 that the input interval is from $x = 0$ to $x = 1$.	Left endpoint? 0 Right endpoint? 1

You are next shown a menu of choices. Always use midpoint rectangles with a limit of sums. Press 4 and ENTER. At the N? prompt, enter 4.	Enter Choice: Left Rect (1) Right Rect (2) Trapezoids (3) Midpt Rect (4) 4 N? 4∎
Record on paper the area estimate for 4 midpoint rectangles. (You should record at least one more decimal place than the number of places needed for the required accuracy. Because Example 1 asks for three decimal-place accuracy, record at least four decimal places of this value.)	Right Rect (2) Trapezoids (3) Midpt Rect (4) 4 N? 4 SUM= .795982305153
Press ENTER to continue the program, and press 1 to choose Change N. Double the number of rectangles by entering N = 8. Record the area approximation (again, to four decimal places).	Change N (1) Change Method (2) Quit (3) 1 N? 8 SUM= .789171732825
Continue in this manner, each time choosing the first option, Change N, and doubling N until a trend is evident.	Change N (1) Change Method (2) Quit (3) 1 N? 256 SUM= .78541918062

NOTE: Intuitively, finding a *trend* means that you are sure what the area approximation, to three decimal places (or whatever accuracy is specified in the problem), will be without having to use larger values of N in the program.

6.4 The Fundamental Theorem

The Fundamental Theorem of Calculus is important because it connects the two main topics in calculus – differentiation and integration. In this section of the *Guide*, we see how to use the TI-86's definite integral function and use it along with the TI-86's numerical derivative to illustrate the Fundamental Theorem in action.

6.4.1 DEFINITE INTEGRAL NOTATION AND CALCULATOR NOTATION Recall that der1 and nDer are the TI-86's derivatives and provide, in most cases, the value of or a good approximation for the instantaneous rate of change of a function when that rate of change exists. As is the case with the TI-86 derivatives, your calculator does not give a formula for accumulation functions. However, it does give an excellent numerical estimate for the definite integral of a function between two specific input values when that integral exists.

The TI's numerical integral is called fnInt, and the correspondence between the mathematical notation $\int_a^b f(x)\,dx$ and the calculator's notation fnInt(*f(x)*, *x*, *a*, *b*) is illustrated in Figure 6.

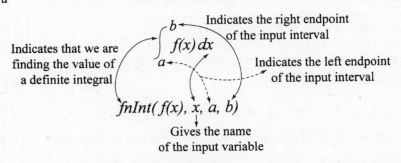

Figure 6

We illustrate the use of the calculator's definite integral function by finding the area from $x = 0$ to $x = 1$ between the function $f(x) = \sqrt{1-x^2}$ and the horizontal axis. Recall that the trend (to three decimal places) found from the limit of sums investigation in Example 1 of Section 6.2 of *Calculus Concepts* was 0.785. (Also see Section 6.5.1 of this *Guide*.)

Type what is shown to the right. Access fnInt with [2nd] [÷] (CALC) [F5] [fnInt]. Note that when you type the function on the home screen, you can use any letter as the input variable. In the y(x)= list, you must use x as the variable name.

NOTE: You can enter f in y1, or any of the other function locations, and refer to it in the fnInt expression as y1 or you can enter the function directly on the home screen as was shown in the last calculation.

6.4.2 THE FUNDAMENTAL THEOREM OF CALCULUS
Intuitively, this theorem tells us that the derivative of an antiderivative of a function is the function itself. Let us view this idea both numerically and graphically. The correct syntax for the TI-86's numerical integrator is

fnInt(function, name of input variable, left endpoint for input, right endpoint for input)

Consider the function $f(t) = 3t^2 + 2t - 5$ and the accumulation function $F(x) = \int_1^x f(t)\, dt$.

The Fundamental Theorem of Calculus tells us that $F'(x) = \frac{d}{dx}\left(\int_1^x f(t)\, dt\right) = f(x)$; that is, $F'(x)$ is f evaluated at x.

Input f in y1 and $F' = $ nDer(fnInt(y1, x, 1, x), x, x) in y2 (remember that the TI-86 requires that you use x as the input variable in the y(x)= list). Access nDer with [2nd] [÷] (CALC) [F2] [nDer] and fnInt with [2nd] [÷] (CALC) [F5] [fnInt]. (Turn off any stat plots that are turned on.)

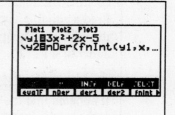

NOTE: The numerical derivative (nDer) is used rather than the function derivative (der1) because fnInt is not one of the pre-programmed functions for which der1 can be calculated.

Press [TABLE] [F2] [TBLST], set Indpnt: to ASK, and press [F1] [TABLE]. Input several different values for x. Other than some occasional round-off error because the calculator is approximating these values, the results are identical.

Find a suitable viewing window such as the one obtained with [GRAPH] [F3] [ZOOM] [F4] [ZSTD]. Without changing the window (that is, draw the graphs by pressing [F5] [GRAPH]), turn off y2 and draw the graph of y1. Then turn off y1 and draw the graph of y2.

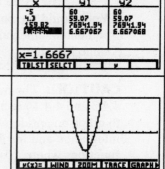

Note that the graph of y2 takes a while to draw. This is because the TI-86 is not only plotting the points, it is also calculating the numerical integral and then the numerical derivative before plotting each point. Turn both y1 and y2 on and draw the graph of both functions. Only one graph is seen in each case.

Enter several other functions in y1 and do not change y2 except possibly for the left endpoint 1 in the fnInt expression. Perform the same explorations as above. Confirm your results with derivative and integral formulas.

6.4.3 DRAWING ANTIDERIVATIVE GRAPHS

Recall when using fnInt($f(x)$, x, a, b) that a and b are, respectively, the lower and upper endpoints of the input interval. Also remember that you do not have to use x as the input variable unless you are graphing the integral or evaluating it using the calculator's table. Unlike when graphing using der1 or nDer, the TI-86 will not graph a general antiderivative; it only draws the graph of a specific accumulation function. Thus, we can use x for the input at the upper endpoint when we want to draw an antiderivative graph, but not for the inputs at both the upper and lower endpoints.

All of the antiderivatives of a specific function differ only by a constant. We explore this idea using the function $f(x) = 3x^2 - 1$ and its general antiderivative $F(x) = x^3 - x + C$. Because we are working with a general antiderivative in this illustration, we do not have a starting point for the accumulation. We therefore choose some value, say 0, to use as the starting point for the accumulation function to illustrate drawing antiderivative graphs. If you choose a different lower limit, your results will differ from those shown below by a constant.

Enter f in y1, fnInt(y1, x, 0, x) in y2, and F in y3, y4, y5, and y6, using a different number for C in each function location. (You can use the values shown to the right or different values.)	
Find a suitable viewing window and graph the functions y1 through y6. The graph to the right was drawn with $-3 \leq x \leq 3$ and $-20 \leq y \leq 20$.	
It appears that the only difference in the graphs of y2 through y6 is the y-axis intercept. But, isn't C the y-axis intercept of each of these antiderivative graphs? Delete y4, y5, and y6. Turn off y1 and change the 1 in y3 to 0.	
Press $\boxed{\text{EXIT}}$ $\boxed{\text{F5}}$ [GRAPH] and draw the graphs of y2 and y3. You should see only one graph. Set the TI-86 TABLE to ASK and enter some values for x. It appears that y2 and y3 are the same function.	

CAUTION: The methods for checking derivative formulas that were discussed in Sections 4.4.3 and 4.4.4 are not valid for checking general antiderivative formulas. Why not? Because to graph an antiderivative using fnInt, you must arbitrarily choose values for the constant of integration and for the input of the lower endpoint. However, for most of the rate-of-change functions where $f(0) = 0$, the calculator's numerical integrator values and your antiderivative formula values should differ by the same constant at every input value where they are defined.

6.5 The Definite Integral

When using the numerical integrator on the home screen, enter fnInt($f(x)$, x, a, b) for a specific function f with input x and specific values of a and b. (Remember that the input variable does not have to be x when on the home screen.) If you prefer, f can be in the y(x)= list and referred to as y1 (or whatever location the function is in) when using fnInt.

6.5.1 EVALUATING A DEFINITE INTEGRAL ON THE HOME SCREEN

We illustrate the use of fnInt with the function that models the rate of change of the average sea level. The rate-of-change data are given in Table 6.43 of Example 3 in Section 6.5 of *Calculus Concepts*.

Time (thousands of years before the present)	−7	−6	−5	−4	−3	−2	−1
Rate of change of average sea level (meters/year)	3.8	2.6	1.0	0.1	−0.6	−0.9	−1.0

Enter the time values in L1 and the rate of change of the average sea level values in L2.

A scatter plot of the data indicates a quadratic function. Find the function and paste it in y1. Draw the function on the scatter plot of the data to confirm that it gives a good fit.	QuadraticReg y=ax²+bx+c n=7 PRegC= {.147619047619 .3595...
Because part *b* of Example 3 asks for the *areas* of the regions above and below the input axis and the function, we must find where the function crosses the axis. You can find this value using the solver (solve y1 = 0) or by using the graph and the *x*-intercept method described in Section 1.1.9 of this *Guide*.	ROOT X=-3.844955338 y=-1E-13

CAUTION: If you round the *x*-intercept value, it will cause whatever you do with this value to not result in as accurate a calculated value as it could be. This situation occurs many times in this and the next several chapters of the text. Also recall that one of the numerical considerations given in Chapter 1 of *Calculus Concepts* is that intermediate calculation values are not to be rounded. For maximum accuracy, store this value in some memory location, say Z, and refer to it as Z in all subsequent calculations.

Whether you use the SOLVER or the *x*-intercept method shown on the calculator screen above, the zero is stored in x. Return to the home screen and store the value in Z (or whatever memory location, except x, that you choose). Do not use the x location because the x value changes whenever you use the TI-86 table, the solver, or trace a graph. (Type Z with [.]; remember that the store command changes the regular cursor to the alphabetic cursor.).

Find the area of the region above the horizontal axis by typing and entering the first expression shown on the right.
 	
Find the area of the region below the horizontal axis by typing and entering the second expression shown on the right. (The negative is used because the region is below the input axis.)
	x→Z
	        -3.84495533758
	fnInt(y1,x,-7,Z)
	        5.40593487597
	-fnInt(y1,x,Z,0)
	        2.93649043153

Part *c* of Example 3 asks you to evaluate $\int_{-7}^{0} y1\, dx$. Find this value by typing and entering the expression shown on the right. Note that the result is not the sum of the two areas – it is their difference.	fnInt(y1,x,-7,0) 2.46944444444

- If you evaluate a definite integral using antiderivative formulas and check your answer with the calculator using fnInt, you might find a slight difference in the last few decimal places. Remember that the TI-86 is evaluating the definite integral using an approximation technique, not an algebraic formula.

6.5.2 EVALUATING A DEFINITE INTEGRAL FROM THE GRAPHICS SCREEN

The value of a definite integral can also be found from the graphics screen. We again illustrate the use of fnInt with the function that models the rate of change of the average sea level. You should practice using the following method by finding the areas of the regions indicated in part *b* of Example 3 in Section 6.5 of the text.

Turn off Plot 1 and have the function modeling the average sea level in y1. (See Section 6.5.1 of this *Guide*.) We want to evaluate $\int_{-7}^{0} y1\, dx$.	
Press [EXIT] [F2] [WIND], set xMin = −7, xMax = 0, and draw the graph of y1 with [F3] [ZOOM] [MORE] [F1] [ZFIT].	
Press [MORE] [F1] [MATH] [F3] [∫f(x)]. The TI-86 asks Lower Limit? Press [(−)] 7 and obtain the screen shown to the right.	
Press [ENTER]. The TI-86 now asks Upper Limit? Press 0 and obtain the screen shown to the right.	
Press [ENTER]. The region between the function and the horizontal axis from the lower limit to the upper limit is shaded. The value of the integral is shown at the bottom of the screen.	

CAUTION: If you type in a value for the lower and/or upper limit (that is, the input at the endpoint) that is not visible on the graphics screen, you will get an error message when you attempt to evaluate the integral. Be certain that these endpoints are included in the interval from xMin to xMax before using this method.

Chapter 7   Analyzing Accumulated Change: Integrals in Action

7.1 Differences of Accumulated Changes

This chapter helps you effectively use your calculator's numerical integrator with various applications pertaining to the accumulation of change. In this first section, we focus on the differences of accumulated changes.

7.1.1 FINDING THE AREA BETWEEN TWO CURVES
The process of finding the area of the region between two functions uses many of the techniques presented in preceding sections. If the two functions intersect, you need to first find the input values of the point(s) of intersection. We show these ideas as they are given in Example 2 of Section 7.1 of *Calculus Concepts*.

Clear all functions from the y(x)= list and turn off the stat plots. Enter the function $f(x) = 0.3x^3 - 3.3x^2 + 9.6x - 5.1$ in y1 and enter the function $g(x) = -0.15x^2 + 2.03x - 5.1$ in y2.	
From Figure 7.10 in the text, we see that the regions under discussion use xMin = 1, xMax = 7, yMin ≈ −3.3, and yMax ≈ 3.3. Press EXIT F2 [WIND], set these values, and then press F5 [GRAPH].	

We next find A and B, the inputs of the two points of intersection of the two functions. (These values will be the limits on the integrals we use to find the areas.) The method we use to find A is the intersection method that was discussed in Section 1.1.9 of this *Guide*.

With the graph on the screen, press MORE F1 [MATH] MORE F3 [ISECT]. The TI-86 asks First curve? Press ENTER to mark y1. The calculator then asks Second curve? Press ENTER to mark y2. At the Guess? prompt, use ◄ to move the cursor near the leftmost point of intersection.	
Press ENTER to find $A \approx 3.724$. To avoid making a mistake copying the x-value and to eliminate as much rounding error as possible, return to the home screen and store the x-value in A with the keystrokes x-VAR STO▶ LOG (A) ENTER.	

You could use this same method to find B, the input value of the rightmost point of intersection. However, to illustrate another way this can be done, we find B using the SOLVER.

The equation we are solving is y1 = y2, but remember that what you enter in the SOLVER must have a constant on one side of the equation. Press ENTER.	

Enter 0 for the right-hand side of the equation (in the second exp: position). Note from the graph that *B* is near 7. Use this or some other value closer to *B* than it is to *A* as your guess in the SOLVER. Press [F5] [SOLVE] to find that $B \approx 6.776$.

As with *A*, to avoid making a mistake copying this *x*-value and to eliminate as much rounding error as possible, return to the home screen and store this *x*-value in *B* with the [x-VAR] [STO▶] [SIN] (B) [ENTER].

Calculate the areas of the three regions enclosed by *f* and *g* (that is, regions R1, R2, and R3 that are shown in Figure 7.10 in the text) as indicated to the right.

NOTE: Each answer should be positive because we are finding areas. If you obtain a negative answer, you probably entered the functions in the wrong order.

If you are in doubt as to which function to enter first in the integral, press [GRAPH] [F5] [GRAPH] and trace the graphs to see which function is on top of the other. Press [▼] to have the cursor jump from one curve to the other and notice the number of the function's position in the top right corner of the screen.

Find $\int_1^7 [f(x) - g(x)]\,dx$ with the instruction shown to the right. Note that the value of this integral is *not* the same as the total area between the two curves.

7.2 Streams in Business and Biology

You will find your calculator very helpful when dealing with streams that are accumulated over finite intervals. Finding either the future or present value of a continuous income stream is simply finding the value of a definite integral. However, the technique used with discrete income streams involves the sequence function that was introduced in Section 2.1.1. The details are presented in Section 7.2.2.

7.2.1 DETERMINING THE FLOW-RATE FUNCTION FOR AN INCOME STREAM
The TI-86 can often help you to find the equation for an income stream flow rate. Note that it is *not* necessary to use the calculator to find such an equation – we present this as a technique to use only if you find writing the equation from the word description difficult. We illustrate these ideas as they are given in Example 1 of Section 7.2 of *Calculus Concepts*.

In Example 1, part *a*, we are told that the business's profit remains constant. Clear lists L1 and L2. In L1 enter two possible input values for the time involved. (You might use different years than the ones shown here.) In L2 enter the amount invested: 10% of the constant profit.

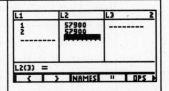

TI-86 Guide

B-77

| You need to remember that a *constant output* means a *linear flow rate*. Fit a linear function to these two data points to find that $R(t) = 57,000$ dollars per year. | |

CAUTION: If you attempt to draw a scatter plot on the calculator, you will get an error message because the TI-86, using the output data in L2, sets yMin = yMax. (You need to draw the scatter plot using paper and pencil.) If you want to see the horizontal line graph on the TI-86, change yMin and yMax so that 57,900 is in between the two values and press F5 [GRAPH].

In Example 1, part *b*, we are told that the business's profit grows by $50,000 each year. The first year's profit (which determines the initial investment at $t = 0$) is $579,000. Reason that if the profit grows by $50,000 each year, the next year's profit will be $579,000 + 50,000 = $629,000. Enter these values in L1 and L2.	
You need to remember that a *constant growth* means a *linear flow rate*. Fit a linear function to these two data points. Now, carefully read the problem once more. Note that only 10% of the profit is invested. Thus, the linear flow rate function is $R(t) = 0.10(50,000t + 579,000)$ dollars per year t years after the first year of business.	
In Example 1, part *c*, we are told that the business's profit grows by 17% each year. The first year's profit (which determines the initial investment at $t = 0$) is $579,000. Reason that if the profit grows by 17% each year, the next year's profit will be $579,000 + 0.17(579,000) = $677,430. Enter these values in L1 and L2.	
You need to remember that a *constant percentage growth* means an *exponential flow rate*. Fit an exponential function to these two data points. Now, carefully read the problem once more. Note that only 10% of the profit is invested. Thus, the exponential flow rate function is $R(t) = 0.10(579,000)(1.17^t)$ dollars per year t years after the first year of business.	

Part *d* of Example 1 gives data that describe the growth of the business's profit. Refer to the material in Section 2.1.4 of this *Guide* to review how to fit a log function to these data points.

NOTE: If you forget which type of growth gives which function, simply use what you are told in the problem and fill in the lists with approximately five data points. Draw a scatter plot and it should be obvious which function to fit to the data.

| For instance, return to Example 1, part *b*, in which we are told that the business's profit grows by $50,000 each year. Note that we assume throughout this section that initial investments are made at time $t = 0$. Simply add $50,000 to the previous year's profit to obtain about five data points. | |

Draw a scatter plot, and observe that the points fall in a line! Note that if you run program DIFF, the first differences are constant at 50,000.

Next, find a linear equation for the data. You should find that the flow rate function is $R(t) = 0.10(50,000t + 579,000)$ dollars per year t years after the first year of business.

7.2.2 FUTURE VALUE OF A DISCRETE INCOME STREAM

The future value of a discrete income stream is found by adding, as d increases, the terms of $R(d)\left(1+\frac{r}{n}\right)^{D-d}$ where $R(d)$ is the value per period of the dth deposit, $100r\%$ is the annual percentage rate at which interest is earned when the interest is compounded once in each deposit period, n is the number of times interest is compounded (and deposits are made) during the year, and D is the total number of deposit periods. It is assumed that initial deposits are made at time $t = 0$ unless it is otherwise stated. We use the situation in Example 4 in Section 7.2 of *Calculus Concepts*:

> When you graduate from college (say, in 3 years), you would like to purchase a car. You have a job and can put $75 into savings each month for this purchase. You choose a money market account that offers an APR of 6.2% compounded quarterly.

Part *a* asks how much will you have deposited in 3 years. No interest is involved in this calculation. There are 12 months in each year and $75 is deposited each month for 3 years. A total of $2700 is deposited.	`12*3*75`
	
	`            2700`

Part *b* of Example 4 asks for the future value of the deposits at the end of 3 years. Because the APR is 6.2% compounded quarterly, $r = 0.062$, $n = 4$, and $D = 3 \text{ years} \cdot 4\frac{\text{deposits}}{\text{year}} = 12$ deposits. A constant $75 each month is deposited, so $R(d) = \$75(3) = \$225$ deposited each quarter. The 3-year future value is given by $\sum_{d=0}^{11} 225\left(1+\frac{0.062}{4}\right)^{12-d}$.

The TI-86 sequence command finds this sum. The syntax for this sequence command is

seq(formula, variable, first value, last value, increment)

When applied to discrete income streams, the *formula* is $R(d)\left(1+\frac{r}{n}\right)^{D-d}$.

For convenience, we enter the formula in y1. You must enclose the exponent in parentheses. Otherwise, the TI-86 will assume that only the 12 is in the exponent.	`Plot1 Plot2 Plot3` `\y1=...+.062/4)^(12-x)`
Return to the home screen and enter seq(y1, x, 0, 11, 1). Recall that the sequence command is accessed by pressing [2nd] [−] (LIST) [F5] [OPS] [MORE] [F3] [seq]. If you want to see the 12 future values in this list, scroll through the list with [▶].	`seq(y1,x,0,11,1)` `(270.608635235 266.4…`

NOTE: When you use the sequence command for discrete income streams, the first value is always 0. The last value is always $D - 1$ because we start counting at 0, not 1. The increment will always be 1 because of the way the formula is designed.

What are the values in the list that results when you use the sequence command? The first value (approximately $270.61) is the 3-year future value of the first 3 months deposits ($225). The second value (approximately $266.48) is the 3-year future value of the second 3 months of deposits ($225), and so forth.

| To find the future value of all the deposits, simply sum the values in the list. To do this, use the sum command to add the values in the list (the previous ANS). The sum command is accessed from the LIST OPS screen with [F1] [sum]. Store this value in F for use in the next part of the problem. | |

Part *c* of Example 4 asks for the present value of the amount you would have to deposit now to achieve the same future value as was found above. We want to find the present value *P*.

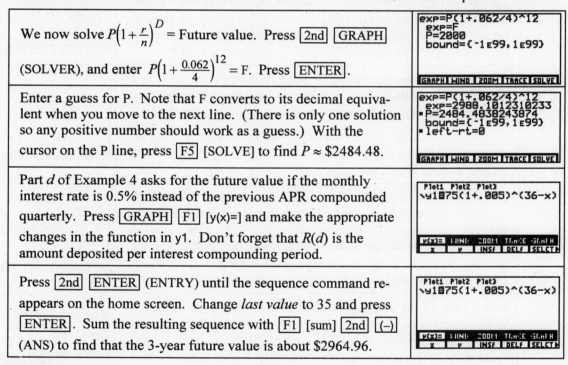

We now solve $P\left(1+\frac{r}{n}\right)^D$ = Future value. Press [2nd] [GRAPH] (SOLVER), and enter $P\left(1+\frac{0.062}{4}\right)^{12}$ = F. Press [ENTER].	
Enter a guess for P. Note that F converts to its decimal equivalent when you move to the next line. (There is only one solution so any positive number should work as a guess.) With the cursor on the P line, press [F5] [SOLVE] to find $P \approx \$2484.48$.	
Part *d* of Example 4 asks for the future value if the monthly interest rate is 0.5% instead of the previous APR compounded quarterly. Press [GRAPH] [F1] [y(x)=] and make the appropriate changes in the function in y1. Don't forget that $R(d)$ is the amount deposited per interest compounding period.	
Press [2nd] [ENTER] (ENTRY) until the sequence command reappears on the home screen. Change *last value* to 35 and press [ENTER]. Sum the resulting sequence with [F1] [sum] [2nd] [(-)] (ANS) to find that the 3-year future value is about $2964.96.	

NOTE: You can use any letter for the variable in the formula if you enter the function on the home screen, but if you are entering the formula in the y(x)= list, use x.

7.3 Integrals in Economics

Consumers' surplus (when defined by a definite integral) and producers' surplus are easy to find using fnInt. You should draw graphs of the demand and supply functions and think of the economic quantities in terms of area to better understand the questions being asked.

7.3.1 CONSUMER ECONOMICS We illustrate how to find the consumers' surplus and other economic quantities when the demand function intersects the input axis as given in Example 2 of Section 7.3 of *Calculus Concepts*:

Suppose the demand for a certain model of minivan in the United States can be described as $D(p) = 14.12(0.933^p) - 0.25$ million minivans when the market price is p thousand dollars per minivan.

We first draw a graph of the demand function. This is not asked for, but it will really help your understanding of the problem. Read the problem to see if there are any clues as to how to set the horizontal view for the graph. The price cannot be negative, so $p \geq 0$. There is no price given in the remainder of the problem, so just guess a value with which to begin.

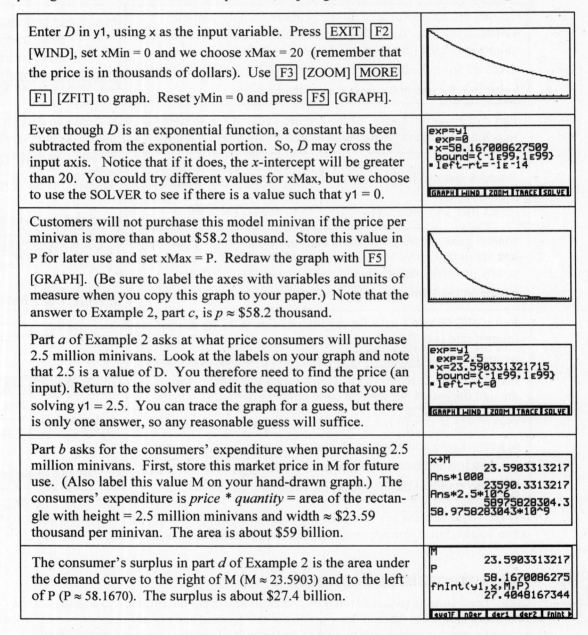

Enter D in y1, using x as the input variable. Press EXIT F2 [WIND], set xMin = 0 and we choose xMax = 20 (remember that the price is in thousands of dollars). Use F3 [ZOOM] MORE F1 [ZFIT] to graph. Reset yMin = 0 and press F5 [GRAPH].	
Even though D is an exponential function, a constant has been subtracted from the exponential portion. So, D may cross the input axis. Notice that if it does, the x-intercept will be greater than 20. You could try different values for xMax, but we choose to use the SOLVER to see if there is a value such that y1 = 0.	
Customers will not purchase this model minivan if the price per minivan is more than about $58.2 thousand. Store this value in P for later use and set xMax = P. Redraw the graph with F5 [GRAPH]. (Be sure to label the axes with variables and units of measure when you copy this graph to your paper.) Note that the answer to Example 2, part c, is $p \approx \$58.2$ thousand.	
Part a of Example 2 asks at what price consumers will purchase 2.5 million minivans. Look at the labels on your graph and note that 2.5 is a value of D. You therefore need to find the price (an input). Return to the solver and edit the equation so that you are solving y1 = 2.5. You can trace the graph for a guess, but there is only one answer, so any reasonable guess will suffice.	
Part b asks for the consumers' expenditure when purchasing 2.5 million minivans. First, store this market price in M for future use. (Also label this value M on your hand-drawn graph.) The consumers' expenditure is *price * quantity* = area of the rectangle with height = 2.5 million minivans and width $\approx \$23.59$ thousand per minivan. The area is about $59 billion.	
The consumer's surplus in part d of Example 2 is the area under the demand curve to the right of M ($M \approx 23.5903$) and to the left of P ($P \approx 58.1670$). The surplus is about $27.4 billion.	

The TI-86 draws the consumers' surplus if you use the methods illustrated in Section 6.5.2. First, reset yMin to -2 to have more room at the bottom of the graph. Reset xMax to 58.2 or an error message comes up with the next instructions. Draw the graph of D with F5 [GRAPH].

With the graph on the screen, press MORE F1 [MATH] F3 [∫f(x)]. At the Lower Limit? prompt, press ALPHA 8 (M) ENTER. At the Upper Limit? prompt, press ALPHA , (P) ENTER. You should then see the screen shown to the right.

CAUTION: Unless the values you enter for the upper and lower limits are visible on (and not on the boundary of) the graphics screen, you will get an error message using the above instructions. If you do as instructed and use P, make sure you used a value slightly larger than P when setting xMax.

7.3.2 PRODUCER ECONOMICS

We illustrate how to find the producers' surplus and other economic quantities as given in Example 4 of Section 7.3 of *Calculus Concepts*:

> Suppose the demand and supply functions for a certain model minivan are described by $D(p) = 14.12(0.933^p) - 0.25$ million minivans and $S(p) = 0$ minivans for $p < 15$ and $S(p) = 0.25p - 3.75$ million minivans for $p \geq 15$ when the market price is p thousand dollars per minivan.

We first draw graphs of the demand and supply functions.

Enter D in y1. Enter S (for $p \geq 15$) in y2 as shown to the right. Use x as the input variable in each function. Recall that the piecewise function and its input each need to be in parentheses. Access the inequality symbol with 2nd 2 (TEST) F5 [≥].	
Press GRAPH F2 [WIND], set xMin = 0 and we choose xMax = 58.167. (This is the approximate value at which $D = 0$, as determined in the previous section.) Use F3 [ZOOM] MORE F1 [ZFIT] to draw the graphs of D and S. If yMin is not 0, set it to 0 and press F5 [GRAPH].	
Part *a* of Example 4 asks for the market equilibrium point. You can find this with the solver (solve y1 – y2 = 0) or by using the intersection method on the graph. We show the latter. Store this x-value in E.	

See Section 7.1.1 for the specifics of using this method. The x- and y-values displayed at the bottom of the screen are the location of the market equilibrium point. The total social gain is producers' surplus + consumers' surplus at the equilibrium price. Shade this area on your graph and then use the graph to write the integrals that give this area:

Social gain = $\int_{15}^{E} S(p)\,dp + \int_{E}^{P} D(p)\,dp$ where $E \approx 24.3996$ and $P \approx 58.1670$. (The value of P was found in Section 7.3.1.)

7.4 Average Value and Average Rates of Change

Average rates of change are computed as discussed in Section 3.1.1 of this *Guide*. When finding an average value, you need to carefully read the question in order to determine which quantity should be integrated. Considering the units of measure in the context can be a great help when trying to determine which function to integrate to find an average value.

7.4.1 AVERAGE VALUE OF A FUNCTION

We illustrate finding an average value with the data in Table 7.18 in Example 1 of Section 7.4 of *Calculus Concepts*:

Time (number of hours after midnight)	7	8	9	10	11	12	13	14	15	16	17	18	19
Temperature (°F)	49	54	58	66	72	76	79	80	80	78	74	69	62

Clear any old data. Delete any functions in the y(x)= list and turn on Plot 1. Enter the data in the above table in lists L1 and L2.

A scatter plot of the data indicates an inflection point (around 9 p.m.) and no limiting values. Fit a cubic function to the data and paste it in y1.

Part *b* of Example 1 asks for the average *temperature* (*i.e.*, the average value of the temperature) between 9 a.m. and 6 p.m. So, integrate the *temperature* between $x = 9$ and $x = 18$ and divide by the length of the interval to find that the answer is about 74.4°F.

7.4.2 GEOMETRIC INTERPRETATION OF AVERAGE VALUE

What does the average value of a function mean in terms of the graph of the function? We continue with Example 1 of Section 7.4 of *Calculus Concepts* by considering the function and average value found in Section 7.4.1. Have the unrounded cubic temperature function in y1.

Enter 74.427229021 in y2. You can either leave yMin at the setting from the scatter plot or change yMin to 0. Press [F5] [GRAPH] to see the graph of the average value and the function.

We illustrate for only the second graph, but what follows is true for both. The area of the rectangle whose height is the average temperature is $(74.427229021 - 0)(18 - 9) \approx 669.845$. The area of the region between the temperature function and the input axis between 9 a.m. and 6 p.m. is this same value.

To find the answer to part *d* of Example 1, enter and evaluate the expression shown to the right. The average rate of change of the temperature from 9 a.m. to 6 p.m. is about 0.98°F/hr.

NOTE: The material in the rest of this *Guide* appears only in the complete version of the text.

7.5 Probability Distributions and Density Functions

Most of the applications of probability distributions and density functions use technology techniques that have already been discussed. Probabilities are areas whose values can be found by integrating the appropriate density function. A cumulative density function is an accumulation function of a probability density function.

Your calculator's numerical integrator is especially useful for finding means and standard deviations of some probability distributions because those integrals often involve expressions for which we have not developed algebraic techniques for finding antiderivatives.

7.5.1 NORMAL PROBABILITIES
The normal density function is the most well known and widely used probability distribution. If you are told that a random variable x has a normal distribution $N(x)$ with mean μ and standard deviation σ, the probability that x is between two real numbers a and b is given by

$$\int_a^b N(x)\,dx = \int_a^b \frac{1}{\sigma\sqrt{2\pi}} e^{\frac{-(x-\mu)^2}{2\sigma^2}} dx$$

We illustrate these ideas with the situation in Example 8 of Section 7.5 of *Calculus Concepts*. In that example, we are given that the distribution of the life of the bulbs, with the life span measured in hundreds of hours, is modeled by a normal density function with $\mu = 900$ hours and $\sigma = 100$ hours. Carefully use parentheses when entering the normal density function.

| The probability that a light bulb lasts between 9 and 10 hundred hours is $$P(9 \le x \le 10) = \int_9^{10} \frac{1}{\sqrt{2\pi}} e^{-(x-9)^2/2}\,dx$$ The value of this integral must be a number between 0 and 1. | | |

7.5.2 VIEWING NORMAL PROBABILITIES
We could have found the probability that the light bulb life is between 900 and 1000 hours graphically by using the CALC menu.

| To draw a graph of the normal density function, note that just about all of the area between this function and the horizontal axis lies within three standard deviations of the mean. Set xMin = 9 − 3 = 6 and xMax = 9 + 3 = 12. Use ZFIT to draw the graph. Reset yMin to −0.1 in preparation for the next step, and press [F5] [GRAPH]. | |
| With the graph on the screen, press [MORE] [F1] [MATH] [F3] [∫f(x)]. At the Lower Limit? prompt, enter 9. At the Upper Limit? prompt, enter 10. The probability is displayed as an area and the value is printed at the bottom of the graphics screen. | |

Chapter 8 Repetitive Change: Cycles and Trigonometry

8.1 Functions of Angles: Sine and Cosine

Before you begin this chapter, go back to the first page of this *Guide* and check the basic setup, the statistical setup, and the window setup. If these are not set as specified in Figures 1, 2, and 3, you will have trouble using your calculator in this chapter. Pay careful attention to the third line in the MODE screen in the basic setup. The Radian/Degree mode setting affects the TI-86's interpretation of the ANGLE menu choices. The calculator's MODE menu should always be set to Radian unless otherwise specified. (Note that calculator instructions for material that is in the *Trigonometry Appendix* at the *Calculus Concepts* Website is on page B-106 of this *Guide*.)

8.1.1 FINDING OUTPUTS OF TRIG FUNCTIONS WITH RADIAN INPUTS It is essential that you have the correct mode set when evaluating trigonometric function outputs. The angle setting in the MODE menu must be Radian for all applications in Chapter 8. We show how to evaluate trig functions as presented in Example 2 of Section 8.1 of *Calculus Concepts*.

Find $\sin \frac{9\pi}{8}$ and $\cos \frac{9\pi}{8}$. Because these angles are in radians, be certain that Radian is chosen in the third line of the MODE screen. The $\boxed{\text{SIN}}$ and $\boxed{\text{COS}}$ keys are above the $\boxed{\text{EE}}$ and $\boxed{\text{(}}$ keys on the TI-86 keyboard.	`sin (9π/8)`
	`   -.382683432365`
	`cos (9π/8)`
	`   -.923879532511`

It is also essential that you use parentheses – the TI-86's order of operations are such that the sine function takes precedence over division. Unless you tell the TI-86 to first divide by using parentheses around $\frac{9\pi}{8}$, it evaluates $(\sin 9) \cdot \frac{\pi}{8}$ and $(\cos 9) \cdot \frac{\pi}{8}$.	`          -.382683432365`
	`cos (9π/8)`
	`          -.923879532511`
	`sin 9π/8`
	`           .161838550706`
	`cos 9π/8`
	`           -.35780001715`

Because the cosine and the sine values are, respectively, the x- and y- coordinates of the point where the terminal side of the angle $\frac{9\pi}{8}$ intersects the unit circle, these values satisfy the equation $x^2 + y^2 = 1$. The fact that $(\cos \theta)^2 + (\sin \theta)^2 = 1$ is true for any angle θ.	`(cos (9π/8))²+(sin (9π/8))²` `                       1`

8.2 Cyclic Functions as Models

We now introduce another model – the sine model. As you might expect, this function should be fit to data that repeatedly varies between alternate extremes. The form of the sine model is given by $f(x) = a \sin (bx + h) + k$ where $|a|$ is the amplitude, b is the frequency (where $b > 0$), $2\pi/b$ is the period, $|h|/b$ is the horizontal shift (to the right if $h < 0$ and to the left if $h > 0$), and k is the vertical shift (up if $k > 0$ and down if $k < 0$). *Note:* The TI-86 uses the c when we use h and d when we use k.

8.2.1 FITTING A SINE MODEL TO DATA Before fitting any model to data, remember that you should construct a scatter plot of the data and observe what pattern the data appear to follow. Example 2 in Chapter 8 asks you to find a sine model for cyclic data with the hours of daylight on the Arctic Circle as a function of the day of the year on which the hours of daylight are measured. (January 1 is day 1.) These data appear in Table 8.3 of *Calculus Concepts*.

Day of the year	−10	81.5	173	264	355	446.5	538	629	720	811.5
Hours of daylight	0	12	24	12	0	12	24	12	0	12

Clear any old data. Delete any functions in the y(x)= list and turn on Plot 1. Enter the data in the above table in lists L1 and L2. Construct a scatter plot of the data. When using the sine regression in the TI-86, it is sometimes necessary to have an estimate of the period of the data.

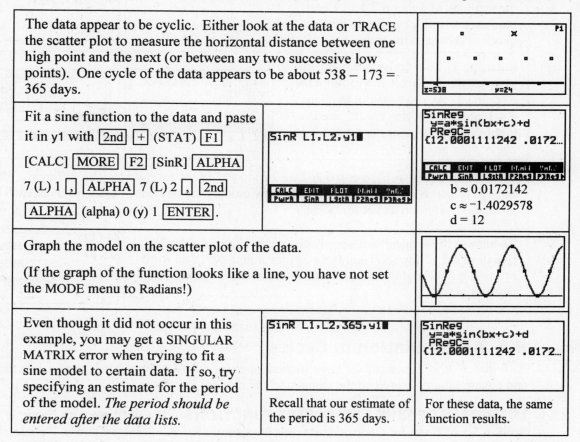

The data appear to be cyclic. Either look at the data or TRACE the scatter plot to measure the horizontal distance between one high point and the next (or between any two successive low points). One cycle of the data appears to be about 538 − 173 = 365 days.

Fit a sine function to the data and paste it in y1 with [2nd] [+] (STAT) [F1] [CALC] [MORE] [F2] [SinR] [ALPHA] 7 (L) 1 [,] [ALPHA] 7 (L) 2 [,] [2nd] [ALPHA] (alpha) 0 (y) 1 [ENTER].

$b \approx 0.0172142$
$c \approx -1.4029578$
$d = 12$

Graph the model on the scatter plot of the data.

(If the graph of the function looks like a line, you have not set the MODE menu to Radians!)

Even though it did not occur in this example, you may get a SINGULAR MATRIX error when trying to fit a sine model to certain data. If so, try specifying an estimate for the period of the model. *The period should be entered after the data lists.*

Recall that our estimate of the period is 365 days.

For these data, the same function results.

NOTE: If you do not think the original function the calculator finds fits the data very well, try specifying a period and see if a better-fitting equation results. It didn't here, but it might with a different set of data. If you still cannot find an equation, you can tell the TI-86 how many times to go through the routine that finds the equation. This number of iterations is 3 if not specified. The number should be typed before L1 when initially finding the equation.

8.3 Rates of Change and Derivatives

All the previous techniques given for other functions also hold for the sine model. You can find intersections, maxima, minima, inflection points, derivatives, integrals, and so forth.

8.3.1 DERIVATIVES OF SINE AND COSINE FUNCTIONS

Evaluate der1 at a particular input to find the value of the derivative of the sine function at that input. We illustrate with Example 1 in Section 8.3 of *Calculus Concepts*:

> The calls for service made to a county sheriff's department in a certain rural/suburban county can be modeled as $c(t) = 2.8 \sin(0.262t + 2.5) + 5.38$ calls during the tth hour after midnight.

Part *a* of this example asks for the average number of calls the county sheriff's department receives each hour. The easiest way to obtain this answer is to remember that the parameter k (d in the TI-86) in the sine function is the average value.

You can also find the average value over one period of the function using the methods of Section 7.4.1. Enter the function c in y1 and type in the quotient shown to the right.	Average Value = $$\frac{\int_0^{2\pi/0.262} c(t)\,dt}{2\pi/0.262 - 0}$$	`fnInt(y1,x,0,2π/.262)`
		                129.021133407
		`Ans/(2π/.262)`
		                         5.38

Enter the TI-86's derivative in y2. Put your answer to part *b*, the formula for $c'(t)$, in y3. As discussed in Sections 4.4.3 and 4.4.4 of this *Guide*, use the graphs of y2 and y3 or the TABLE to check your answer.	Plot1 Plot2 Plot3 \y1=2.8sin (.262x+2.... \y2=der1(y1,x,x) \y3=.7336cos (.262x+...	(table showing x, y2, y3 values)

We entered $c'(x) = 0.7336 \cos(0.262x + 2.5)$ in y3. Your formula may be different. We now go to part c of Example 1 that asks how quickly the number of calls received each hour is changing at noon and at midnight.

To answer these questions, simply evaluate y2 (or your derivative in y3) at 12 for noon and 0 (or 24) for midnight. (You were not told if "midnight" refers to the initial time or 24 hours after that initial time.)	`y2(12)`
	           .588774171694
	`y2(0)`
	         -.587718956365
	`y2(24)`
	         -.589825974892

8.5 Accumulation in Cycles

As with the other functions we have studied, applications of accumulated change with the sine and cosine functions involve the calculator's numerical integrator fnInt.

8.5.1 INTEGRALS OF SINE AND COSINE FUNCTIONS

We illustrate with Example 1 in Section 8.5 of *Calculus Concepts* that uses the rate of change of temperature in Philadelphia on August 27, 1993: $t(x) = 2.733 \cos(0.285x - 2.93)$ °F per hour, x hours after midnight.

Enter the function t in y1. Find the accumulated change in the temperature between 9 a.m. and 3 p.m. using fnInt.		

The temperature increased by about 13°F between 9 a.m. and 3 p.m.

Chapter 9 Ingredients of Multivariable Change: Models, Graphs, Rates

9.1 Cross-Sectional Models and Multivariable Functions

For a multivariable function with two input variables, obtain a cross-sectional model by entering the data in lists L1 and L2 and then fitting the appropriate function as indicated in previous chapters of this *Guide*. Unless you are told otherwise, we assume that the data are given in a table with the values of the *first* input variable listed *horizontally* across the top of the table and the values of the *second* input variable listed *vertically* down the left side of the table.

9.1.1 FINDING A CROSS-SECTIONAL MODEL FROM DATA (HOLDING THE FIRST INPUT VARIABLE CONSTANT)
Using the elevation data in Table 9.2 of Section 9.1 in *Calculus Concepts*, find the cross-sectional model $E(0.8, n)$ as described below. Remember that "rows" go from left to right horizontally and "columns" go from top to bottom vertically.

n (miles)	1.5	1.4	1.3	1.2	1.1	1.0	0.9	0.8
Elevation (feet above sea level)	797.6	798.1	798.5	798.9	799.2	799.5	799.7	799.9
n (miles)	0.7	0.6	0.5	0.4	0.3	0.2	0.1	0.0
Elevation (feet above sea level)	800.0	800.1	800.1	800.1	800.0	799.9	799.7	799.5

So, refer for a moment to Table 9.1. When you are asked to find $E(0.8, n)$, e is constant at 0.8 and n varies. Thus, enter the values for n appearing on the left side of the table (vertically) in L1 and the elevations E obtained in the $e = 0.8$ column of Table 9.1 in L2. *The outputs that you enter in L2 will always be in the main body of any multivariable data table given in this text.*

After entering the data, clear any previously entered functions from the y(x)= list, and turn on Plot1. Draw a scatter plot of the data with F3 [ZOOM] MORE F5 [ZDATA].		
The data appear to be quadratic. Fit a quadratic function and copy it to the y(x)= list. (Refer to Section 2.4.2 of this *Guide*.) Overdraw the graph of the function on the scatter plot with GRAPH F5 [GRAPH].		

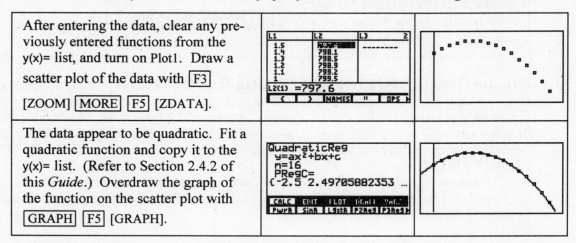

CAUTION: Because you will often be asked to find several different cross-sectional models using the same data table, calling different variables by the same names x and y would be very confusing. It is very important that you call the variables by the names that have been assigned in the problem. Remember that when finding or graphing a function, the TI-86 always calls the input variable x and the output variable y. When working with multivariable functions, you must translate the calculator's equation y1 ≈ −2.5x^2 + 2.497 x + 799.490 into the symbols that are used in the application. You should write the cross-sectional function as $E(0.8, n) = -2.5n^2 + 2.497n + 799.490$. Don't forget to completely describe (including units) all of the variables.

9.1.2 FINDING A CROSS-SECTIONAL MODEL FROM DATA (HOLDING THE SECOND INPUT VARIABLE CONSTANT) The only difference in this and the previous section of this *Guide* is that the second input, instead of the first, is held constant. We illustrate using the situation in Example 1 of Section 9.1 in *Calculus Concepts*. Refer to Table 9.1. Because we are asked to find the cross-sectional model $E(e, 0.6)$, $n = 0.6$ and the inputs are the values of e that are across the top of the table. Enter these values in L1. (See below for a shortcut.) The outputs are the elevations E obtained in the $n = 0.6$ mile row in Table 9.1. Enter these in L2.

NOTE: You may find it helpful to place a piece of paper or a ruler under the row (or to the right of the column) in which the data appear to help avoid typing an incorrect value.

Because the input values for this function are the same as the input values in the last section of this *Guide* (but in reverse order) you can sort L1 in ascending order to avoid re-entering the data. On the home screen, press 2nd – (LIST) F5 [OPS] F2 [sortA] ALPHA 7 (L) 1 ENTER. Store the sorted list in L1 and then enter the output in L2.		
After entering the data, clear any functions from the y(x)= list, and turn on Plot1. Draw a scatter plot of the data.		
There is an inflection point and no evidence of limiting values, so the data appear to be cubic. Fit a cubic function and copy it to the y(x)= list. (Refer to Section 2.4.3 of this *Guide*.) Draw the cubic function on the scatter plot with GRAPH.		

9.1.3 EVALUATING OUTPUTS OF MULTIVARIABLE FUNCTIONS As is the case with single-variable functions, outputs of multivariable functions are found by evaluating the function at the given values of the input variables. The main difference is that you usually will not be using x as the input variable symbol. One way to find multivariable function outputs is to evaluate them on the home screen. We illustrate with the investment function in Example 2 of Section 9.1 in *Calculus Concepts*.

The answer to part *a* of Example 2, as derived from the compound interest formula, uses the formula for the accumulated amount of an investment of P dollars for t years in an account paying 6% interest compounded quarterly:

$$A(P, t) = P\left(1 + \frac{0.06}{4}\right)^{4t} \text{ dollars}$$

When 10 is substituted for t, the cross-sectional function becomes $A(P, 10) \approx 1.814018409P$.

Part *b* of Example 2 asks for a model for the data in the row of Table 9.4 that corresponds to 10 years. Thus, we want to find $A(P, 10)$. Enter the data from the table in L1 and L2.

A scatter plot of the data appears to be linear. Program DIFF confirms that a linear function is appropriate because the first differences are constant. Fit a linear function and paste it in y1.

Notice that the cross-sectional functions in parts *a* and *b* are slightly different because the data in Table 9.4 have been rounded to the nearest cent. Part *c* of Example 2 asks for $A(5300, 10)$. Even though it is simplest here to substitute 5300 for P in $A(P, 10) \approx 1.814018409P$, we return to the original function to illustrate evaluating multivariable formulas on the calculator.

To find the output on the home screen, type the formula for $A(P, t)$, substituting $P = 5300$ and $t = 10$. Press ENTER.

Again, be warned that you must carefully use the correct placement of parentheses. Check your result with the values in Table 9.4 in the text to see if $9614.30 is a reasonable amount.

Even though it is not necessary in this example, you may encounter activities in this section in which you need to evaluate a multivariable function at several different inputs. You could use what is shown above, but there are easier methods than entering each one individually. You will also use the techniques shown below in later sections of this chapter. When evaluating a multivariable function at several different input values, you may find it more convenient to enter the multivariable function in the graphing list.

Clear any previously-entered equations. Enter in y1 the function $A(P, t) = P\left(1 + \frac{0.06}{4}\right)^{4t}$ with the keystrokes ALPHA , (P) ((1 + .06 ÷ 4)) ^ ((4 ALPHA − (T))).

Return to the home screen and input the values $P = 5300$ and $t = 10$ with the keystrokes 5300 STO▶ , (P) 2nd . (:) ALPHA 10 STO▶ − (T) ENTER. Evaluate y1 at these inputs with 2nd ALPHA (alpha) 0 (y) 1 ENTER.

To evaluate y1 at other inputs, press 2nd ENTER (ENTRY) twice to recall the storing instruction. Change the values and press ENTER. Then press 2nd ENTER (ENTRY) twice to recall y1 and press ENTER.

CAUTION: It is very important to note at this point that while we have previously used x as the input variable when entering functions in the y(x)= list, we do *not* follow this rule when we *evaluate* functions with more than one input variable. However, realize that we should not graph y1 nor use the TABLE. If you attempt to graph the current y1 = $A(P, t)$ or use the table, you will see that the calculator considers this y1 a constant. (Check it out and see that the graph is a horizontal line at about 10467.1 and that all values in the table are about 10467.1.)

9.2 Contour Graphs

Because any program that we might use to graph[1] a three-dimensional function would be fairly involved and take a long time to execute, we do not graph three-dimensional functions. Instead, we discover information about three-dimensional graphs using their associated contour curves.

9.2.1 SKETCHING CONTOUR CURVES   When given a multivariable function with two input variables, you can draw contour graphs using the three-step process described below. We illustrate with the function that gives the monthly payments required to pay off a loan of A thousand dollars over a period of t years at 7% interest. This function, M, appears in Example 1 of Section 9.2 of *Calculus Concepts*:

$$M(t, A) = \frac{5.833333A}{1 - 0.932583^t} \text{ dollars}$$

Step 1: We are told that the monthly payment on the loan is $520. Set $M(t, A) = 520$. Because we will use the solve routine to find the value of A at various values of t, write the function as $M(t, A) - 520 = 0$.

Step 2: Choose values for t and solve for A to obtain points on the $520 constant-contour curve. Obtain guesses for the values of t and A from the table of loan amounts in Figure 9.15 of the text.

Enter the $M(t, A)$ formula in y1. Be sure that you use the letters A and T and not x. (We are only using y1 as a "holding place" for the formula.) Go to the SOLVER and enter y1 in the eqn: position. Press [ENTER].		
Enter 520 (the right-hand side of the equation) in the exp: position. Use [▼] to move to the T line and type 12. Press [▲] to move the cursor to the A line and enter a guess. The cursor must be on the line of the unknown variable when you press [F5] [SOLVE].		
Enter another value for t, say $t = 16$ years. Move the cursor to the A line and press [F5] [SOLVE] to find $A \approx$ $59.963 thousand.		
Repeat the procedure for $t = 20$, $t = 24$, $t = 28$, and $t = 30$. Make a table of the values of t and A as you find them.		

---

[1] There are many TI-86 programs, including one that will graph a multivariable function with two input variables, available at the web site with address http://education.ti.com/product/tech/86/down/download.html.

Step 3: Plot the points obtained in Step 2 with pencil and paper. You need to find as many points as it takes to see the pattern the points are indicating when you plot them. Connect the points with a smooth curve.

You need a function to draw a contour graph using the above method. Even though there may be several functions that seem to fit the data points obtained in Step 2, their use would be misleading because the real best-fit function can only be determined by substituting the appropriate values in a multivariable function. The focus of this section is to use contour graphs to study the relationships between input variables, not to find the equation of a function to fit a contour curve. Thus, we always sketch the contours on paper rather than with the TI-86.

9.3 Partial Rates of Change

When you hold all but one of the input variables in a multivariable function constant, you are actually looking at a function of one input variable. Thus, every technique we discussed previously can be used. In particular, the calculator's derivatives der1 and nDer can be used to find partial rates of change at specific values of the varying input variable. Although your TI-86 does not give formulas for derivatives, you can use it as discussed in Sections 4.4.4 and/or 4.4.4 of this *Guide* to check your algebraic formula for a partial derivative.

9.3.1 NUMERICALLY CHECKING PARTIAL DERIVATIVE FORMULAS As mentioned in Chapter 4, the basic concept in checking your algebraically-found partial derivative formula is that your formula and the calculator's formula computed with der1 or nDer should have the same outputs when each is evaluated at several different randomly-chosen inputs. You can use the methods in Section 9.1.3 of this *Guide* to evaluate each derivative formula at several different inputs and determine if the same numerical values are obtained from each formula. We illustrate these ideas by checking the answers for the partial derivative formulas found in parts *b* and *d* of Example 1 for this investment function:

The accumulated value of an investment of *P* dollars over *t* years at an APR of 6% compounded quarterly is $A(P, t) = P(1.061363551^t)$.

Recall that the syntax for the calculator's derivative that we use whenever possible is

der1(function, symbol for input variable, point at which the derivative is evaluated)

Enter the function *A*, using the letters P and T that appear in the formula, in y1. Part *b* of Example 1 asks for a formula for $\partial A/\partial t$, so enter your formula (which may not be the same as the one that is shown to the right) in y2. Enter the TI-86's derivative, using T as the changing input, in y3.	Plot1 Plot2 Plot3 \y1=P(1.061363551^T) \y2=P ln(1.061363551… \y3=der1(y1,T,T) P INSf DELf SEL evalF | nDer | der1 | der2 | fnInt ▶

Store a value in T and call up y2 and y3. Do this for some different inputs. If the outputs of y2 and y3 are the same, your answer in y2 is *probably* correct.	5→T:y2
	          521.372863237
	y3
	          521.372863237
	10→T:y2
	          702.21365237
		702.21365237
		y3
		          702.21365237
		25→T:y2
		          1715.66229132
		y3
		          1715.66229132

NOTE: Remember that [2nd] [ENTER] (ENTRY) recalls previously-entered statements so you can save time. Also recall that [2nd] [.] (:) joins two or more statements together. Only the value of the last statement is printed, so do not join all three together with this symbol.

You may find it more convenient to numerically check your answer using the TABLE. If so, you must remember that when using the TABLE, the TI-86 considers x as the variable that is changing. When finding a partial derivative formula, all other variables are held constant except the one that is changing. So, to use the TABLE (or draw a graph), just store values in all constants and call the changing variable x. Then, proceed according to the directions given in Chapter 4. We illustrate this idea with part *d* of Example 1.

Have the function A, using the letters P and T that appear in the formula, in y1. Because P is the variable that is changing in $\partial A/\partial P$, replace P with x. Turn off y1.
 	
Enter your formula for $\partial A/\partial P$ (which may not be the same as the one that is shown to the right) in y2. Enter the TI-86's derivative, using x as the changing input, in y3.
	`\y1=x(1.061363551^T)`
	`\y2=(1.061363551^T)`
	`\y3=der1(y1,x,x)`

Return to the home screen. Because T is constant, enter any reasonable value, except 0, in T. (You could repeat what comes next for T = 2, 10, and 37 as shown in part *c* of Example 1.) (If you need additional instructions on using the TABLE, see Section 1.1.6 of this *Guide*.)	`2→T` `2`
Have TBLSET set to ASK in the Indpnt: location. Go to the TABLE and input some reasonable values for $t = x$. If the values shown for y2 and y3 are the same, your formula in y2 for $\partial A/\partial P$ is *probably* correct.	TABLE SETUP TblStart=0 △Tbl=1 Indpnt: Auto **Ask** / x, y2, y3 columns showing 1.12649... values y2=1.1264925873913

Did you wonder why the values in y2 and y3 are the same for all values of T? (It is because the function in y1 is a linear function with a constant slope. Remember that T is a constant because P is the changing variable in $\partial A/\partial P$.)

NOTE: If you have an unrounded function that was fit to data in y1 and you use the rounded function to compute your partial derivative, the y2 and y3 columns will be slightly different.

9.3.2 VISUALIZING AND ESTIMATING PARTIAL RATES OF CHANGE A partial rate of change of a multivariable function (evaluated at a specific point) is the slope of the line tangent to the graph of a cross-sectional model at a given location. We illustrate this concept in this section and in Section 9.3.3 using the Missouri farmland cross-sectional equations for elevation: $E(0.8, n)$ and $E(e, 0.6)$. It would be best to use the unrounded functions that were found in Sections 9.1.1 and 9.1.2 of this *Guide*. However, here we use the rounded functions rather than re-enter all the data.

Enter $-2.5n^2 + 2.497n + 799.490 = E(0.8, n)$ in y1. Because we are going to graph this function, use x, not *n*, as the input variable. Press EXIT F2 [WIND] and set values such as those shown to the right.	`\y1=-2.5x²+2.497x+79...`	WINDOW xMin=0 xMax=1.5 xScl=1 yMin=795 yMax=804 ↓yScl=1

The window settings can be obtained by drawing a scatter plot of the data used to obtain the function $E(0.8, n)$ or by looking at the $e = 0.8$ column in Table 9.1 in *Calculus Concepts*.

You can draw the line tangent to the graph at $n = x = 0.6$ from either the home screen or the graphics screen. (See Section 3.2.2 of this *Guide* for an explanation of both methods.) We use the home screen in this section and the graphics screen in the next section.

Press [F5] [GRAPH] to draw the graph of y1. Draw the tangent line to $E(0.8, n)$ at $n = 0.6$ with [MORE] [F2] [DRAW] [MORE] [MORE] [MORE] [F2] [TanLn], [2nd] [ALPHA] (alpha) 0 (y) 1 [,] [.] 6 [)]	Press [ENTER] to see the tangent line.	
You can obtain an *estimate* of the tangent line slope by pressing [MORE] [F1] [MATH] [F2] [dy/dx]. Use [▶] or [◀] to move the cursor as close as possible to $n = x = 0.6$. Press [ENTER].	0.6 is not a trace value.	This is just an estimate.

We could have obtained a much better estimate than the one given above if we had found the slope of the tangent line at $x = 0.6$ rather than a value close to that number. The TI-86 allows you to do this on the home screen or graphically. Here we illustrate the graphical method.

Have $E(0.8, n)$, using x as the input variable, in y1. Draw the tangent line to $E(0.8, n)$ at $n = 0.6$ by first drawing the graph of the function with [F5] [GRAPH]. On the graph screen, press [MORE] [F1] [MATH] [MORE] [MORE] [F1] [TANLN] [.] 6 [ENTER].	

The slope calculated by der1 is shown. So, $\dfrac{dE(0.8, n)}{dn}$ at $n = 0.6$ is -0.503 foot/mile.

9.3.3 FINDING PARTIAL RATES OF CHANGE USING CROSS-SECTIONAL MODELS

The procedures described in the last section required that you use x as the input variable to find partial rates of change. The process of replacing other variables with x can be confusing. You can avoid having to do this replacement when you use the TI-86 derivative on the home screen to evaluate partial rates of change at specific input values. We illustrate with the cross-sectional function $E(e, 0.6) = -10.124e^3 + 21.347e^2 - 13.972e + 802.809$ feet above sea level that is given in Example 2 of Section 9.3 in *Calculus Concepts*. We also demonstrate this method with the cross-sectional function $E(0.8, n)$ used in the last section of this *Guide*.

Have $E(0.8, n) = -2.5n^2 + 2.497n + 799.490$, using N as the input variable, in y1. Enter $E(e, 0.6)$, using E as the input variable, in y2. (Remember that because we are not using x as the input variable, we should not draw a graph or use the TABLE.)	

NOTE: These functions could be entered on the home screen instead of in the y(x)= list. We use y1 and y2 in the y(x)= list because they are convenient locations to hold the equations.

Tell the TI-86 the point at which the derivative is to be evaluated by storing the values. Then, on the home screen, find $\dfrac{dE(0.8,\, n)}{dn}$ evaluated at $n = 0.6$ with the derivative der1.

Find $\dfrac{dE(e,\, 0.6)}{de}$ evaluated at $e = 0.8$ to be about 0.745 foot per mile. (Always remember to attach units of measure to the numerical values when writing your answer.)

NOTE: It is not necessary to again store the values for N and E unless for some reason they have been changed from the values stored previously.

If you want to see the line tangent to the graph of $E(e,\, 0.6)$ at $e = 0.8$, first change every E in y2 to x and turn off y1. Draw the graph of y2 and use the TANLN instruction in the DRAW MATH menu.

9.4 Compensating for Change

As you have just seen, the TI-86 closely estimates numerical values of partial derivatives with nDer and finds the numerical value of the derivative with its der1 function. This technique can also be very beneficial and help you eliminate many potential calculation mistakes when you find the rate of change of one input variable with respect to another input variable (that is, the slope of the tangent line) at a point on a contour curve.

9.4.1 EVALUATING PARTIAL DERIVATIVES OF MULTIVARIABLE FUNCTIONS The last few sections of this *Guide* indicate how to estimate and evaluate partial derivatives using cross-sectional models. The TI-86 evaluates partial derivatives calculated directly from multivariable function formulas using the same procedures. The most important thing to remember is that you must supply the name of the input variable that is changing and the values at which the partial derivative is evaluated. We illustrate using the body-mass index function that is in Example 1 of Section 9.4 of *Calculus Concepts*:

A person's body-mass index is given by $B(h,\, w) = \dfrac{0.4536w}{0.00064516 h^2}$ where h is the person's height in inches and w is the person's weight in pounds. We first find B_h and B_w at a specific height and weight and then use those values in the next section of this *Guide* to find the value of the derivative $\dfrac{dw}{dh}$ at that particular height and weight. The person in this example is 5 feet 7 inches tall and weighs 129 pounds.

Enter B in the y1 location of the y(x)= list, using the letters H and W for the input variables. Next, store the values of H and W at the given point.

CAUTION: Forgetting to store the values of the point at which the derivative is to be evaluated is a common mistake made when using this method. The TI-86 derivative instructions specify the value at only one input, and you must tell the calculator the values of all inputs.

Find the value of B_h at $h = 67$ and $w = 129$ by evaluating the expression shown to the right. The symbol B_h (or $\partial B/\partial h$) tells you that h is varying and w is constant, so h is the variable you type in the numerical derivative. Enter the value of h after the letter. Store this value in N for use in the next section of this *Guide*.	`der1(y1,H,67)`
	`      -.60311608406`
	`Ans→N`
	`      -.60311608406`
	`[evalF][nDer][der1][der2][fnInt]`

Find the value of B_w at $h = 67$ and $w = 129$ by evaluating the expression shown to the right. The symbol B_w (or $\partial B/\partial w$) tells you that w is varying and h is constant, so w is the variable you type in the numerical derivative. Enter the value of w after the letter. Store this value in D for use in the next section of this *Guide*.	`der1(y1,W,129)`
	`       .156623169116`
	`Ans→D`
	`       .156623169116`
	`[evalF][nDer][der1][der2][fnInt]`

9.4.2 FINDING THE SLOPE OF A LINE TANGENT TO A CONTOUR CURVE

We continue the previous illustration with the body-mass index function in Example 1 of Section 9.4 of *Calculus Concepts*. Part *a* of Example 1 asks for $\frac{dw}{dh}$ at the point (67, 129) on the contour curve corresponding to the person's current body-mass index. The formula is $\frac{dw}{dh} = \frac{-B_h}{B_w}$.

An easy way to remember this formula is that whatever variable is in the numerator of the derivative (in this case, w) is the same variable that appears as the changing variable in the denominator of the slope formula. This is why we stored B_w as D (for denominator) and B_h as N (for numerator). Don't forget to put a minus sign in front of the numerator.

In the previous section, we stored B_h as N and B_w as D. So, $\frac{dw}{dh} = \frac{-B_h}{B_w} = \text{-N} \div \text{D}$. (Storing these values also avoids round-off error.) The rate of change is about 3.85 pounds per inch.	`-N/D` `       3.85074626866`

9.4.3 COMPENSATING FOR CHANGE

When one input of a two-variable multivariable function changes by a small amount, the value of the function is no longer the same as it was before the change. The methods illustrated below show how to determine the amount by which the other input must change so that the output of the function remains at the value it was before any changes were made. We again continue the previous illustration with the body-mass index function and part *b* of Example 1 of Section 9.4 of *Calculus Concepts*.

To estimate the change in weight needed to compensate for growths of 0.5 inch, 1 inch, and 2 inches if the person's body-mass index is to remain constant, you need to find $$\Delta w \approx \frac{dw}{dh}(\Delta h)$$ at the given values of Δh. Again, to avoid rounding error, it is easiest to store the slope of the tangent line to some location, say T, and use the unrounded value to calculate.	`                 3.85074626866`
	`Ans→T`
	`                 3.85074626866`
	`T*.5`
	`                 1.92537313433`
	`T*2`
	`                 7.70149253731`
	
	These are changes in weight, so the units that should be attached are *pounds*.

Chapter 10 Analyzing Multivariable Change: Optimization

10.2 Multivariable Optimization

As you might expect, multivariable optimization techniques that you use with your TI-86 are very similar to those that were discussed in Chapter 5. The basic difference is that the algebra required to get the expression that comes from solving a system of equations with several unknowns reduced to one equation in one unknown is sometimes difficult. However, once your equation is of that form, all the optimization procedures are basically the same as those that were discussed previously.

10.2.1 FINDING CRITICAL POINTS USING MATRICES Critical points for a multivariable function are points at which maxima, minima, or saddle points occur. To find critical point(s) of a smooth, continuous multivariable function, we first find the partial derivatives with respect to each of the input variables and then set the partial derivatives equal to zero. This gives a *system of equations* to be solved. Here we illustrate solving this system using an orderly array of numbers that is called a *matrix*. The TI-86 simplifies this solving process using matrices with its *simultaneous equation solver* accessed with 2nd TABLE (SIMULT).

WARNING: This solution method applies only to *linear* systems of equations. That is, the system of equations should not have any variable appearing to a power higher than 1 and should not contain a product of any variables. If the system is not linear, you must use algebraic solution methods to solve the system. Using your TI-86 with the algebraic solution method is illustrated in Section 10.2.2 of this *Guide*.

Consider the function $P(h, c)$ that gives the percentage of sugar converted to olestra when the ration of peanut oil to sugar is 10 to 1. This function is given at the beginning of Section 10.2 of *Calculus Concepts*:

$$P(h, c) = -1924.3916 + 24.5402h + 25.5695c - 0.5927h^2 - 0.0860c^2 - 0.0810hc$$

The system of equations derived from the partial derivatives of P are

$$P_c = 25.5695 - 0.172c - 0.081h = 0$$
$$P_h = 24.5402 - 1.1854h - 0.081c = 0$$

In order to enter the correct values into the simultaneous equation solver, <u>you MUST write the equations so that the constant terms are on the right-hand side of the equations and the coefficients of the input variables occupy the same positions on the left-hand side of each equation</u>:

$$-0.081h - 0.172c = -25.5695 \qquad [1]$$
$$-1.1854h - 0.081c = -24.5402 \qquad [2]$$

The simultaneous equation solver is accessed with 2nd TABLE (SIMULT).

Enter 2 for the number of equations. You next see a screen with the letters a, b, and x with some subscripts.	SIMULT Number=2∎	a1,1x1+a1,2x2=b1 a1,1=∎ a1,2= b1= PREV | NEXT | CLRa | | SOLVE

The subscripts give the row and column that the coefficients in the equations appear. The *a*s represents the coefficients on the left-hand side of the equation and *b* represents the constant term on the right-hand side of the equation. The first value in the subscript stands for the equation and the second number stands for the column in which the variable appears. For instance, "$a_{1,2}$" refers to equation 1 and the coefficient of the second variable, c. Note that x_1 refers to the input variable in the first position (our h) and x_2 refers to the input variable in the second position (our c).

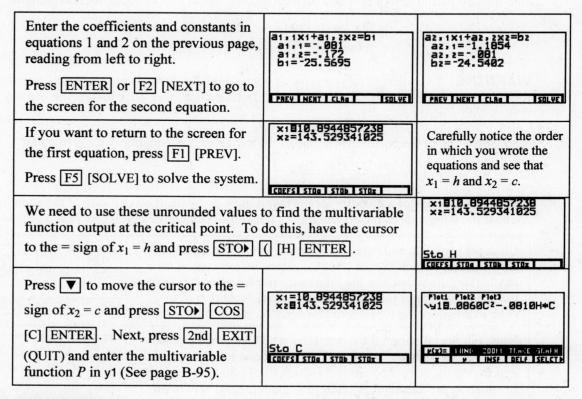

WARNING: Remember that you need to enter a times sign ([×]) between two or more variables whose names are written together so that the TI-86 will recognize the two individual variables rather than think that there is a new input called "HC".

We have found that $h \approx 10.9$ hours and $c \approx 143.5°C$. On the home screen find $P(h, c)$ by typing y1 [ENTER]	

10.2.2 FINDING CRITICAL POINTS USING ALGEBRA AND THE SOLVER

We begin the process of finding critical points of a smooth, continuous multivariable function by using derivative formulas to find the partial derivative with respect to each input variable and setting these partial derivatives each equal to 0. We next use algebraic methods to obtain one equation in one unknown input. Then, you can use your calculator's solve routine to obtain the solution to that equation. This method of solution works for all types of equations. We illustrate these ideas with the function in Example 2 of Section 10.2. (Note that the partial derivative equations in Example 2 are not linear; thus, the simultaneous equation solver can not be used.)

The total daily intake of organic matter required by a beef cow grazing the Northern Great Plains rangeland can be modeled by

$$I(s, m) = 8.61967 - 1.244s + 0.0897s^2 - 0.20988m + 0.035947m^2 + 0.214915sm$$

kg per day when the cow produces m kilograms of milk per day and s is a number between -4 and 4 indicating the size of the cow.

Find the two partial derivatives and set each of them equal to 0 to obtain these equations:

I_s: $\quad -1.244 + 0.1794s + 0.214915m = 0$ $\qquad$ [3]

I_m: $\quad -0.20988 + 0.071894m + 0.214915s = 0$ $\qquad$ [4]

WARNING: Everything that you do with your calculator depends on the partial derivative formulas that you find using derivative rules. Double check this work before you use any of these solution methods.

Next, solve one of the equations for one of the variables, say equation 3 for s, to obtain

$$s = \frac{1.244 - 0.214915m}{0.1794}$$

Recall that to *solve* for a quantity means that it must be by itself on one side of the equation and that the other side of the equation does not contain that letter. Now, let your TI-86 work.

Clear the y(x)= list, and enter the function I in y1. (Remember to enter the last term as 0.214915S*M.) Enter the expression for s in y2. (Be sure to enclose the numerator in parentheses.) Type the left-hand side of the *other* equation (here, equation 4) in y3. (The expression you type should be equal to zero.)	Plot1 Plot2 Plot3 \y1■8.61967-1.244S+... \y2■(1.244-.214915M)... \y3■-.20988+.071894M...
Now, put the cursor on the letter S in the y3 location. Replace *every* S in y3 with y2. What you have just done is substitute the expression for S from equation 3 (which the TI-86 knows as y2) in equation 4! The expression in y3 is the left-hand side of an equation that equals 0 and contains only one variable, namely m.	Plot1 Plot2 Plot3 \y1■8.61967-1.244S+... \y2■(1.244-.214915M)... \y3■...1894M+.2149152■
The next step is to use the SOLVER to solve the equation y3 = 0. Try several different guesses and see that, for this equation, they all result in the same solution. (See Section 1.1.7 if you need to review using the SOLVER.)	eqn:y3■ $\qquad$ exp=y3 exp=0 ■M=6.8998840512347 bound={-1E99,1E99} ■left-rt=-2E-14

NOTE: You need to examine the equation in y3 and see what type of function it represents. In this case, y3 contains m to no power higher than one, so it is a linear equation and has only one root. If y3 contains the variable squared, the equation is quadratic and you need to try different guesses because there could be two solutions, and so forth.

Return to the home screen and press ALPHA 8 ENTER to check that the TI-86 knows that the answer from the SOLVER is a value of M. Next, find s by evaluating y2. The TI-86 does not know that the output of y2 equals S, so be sure to store this value in S.	M $\qquad$ 6.89988405123 y2 $\qquad$ -1.33159744075 Ans→S $\qquad$ -1.33159744075

The last thing to do is to find the output $I(s, m)$ at the current values of s and m. To do this, call up y1 on the home screen.	y2 6.89988405123 Ans→S -1.33159744075 -1.33159744075 y1 8.72384977581

The critical point has coordinates $s \approx -1.33$, $m \approx 6.90$ kg of milk, and $I \approx 8.72$ kg of organic matter eaten each day.

10.2.3 CLASSIFYING CRITICAL POINTS USING THE DETERMINANT TEST

Once you find one or more critical points, the next step is to classify them as points at which a maximum, a minimum, or a saddle point occurs. The Determinant Test often will give the answer. Also, because this test uses derivatives, the calculator's derivative der1 can help.

We illustrate with the critical point that was found for Example 2. To use the Determinant Test, we need to calculate the four second partial derivatives of I and evaluate them at the critical point values of S and M.

Enter the function I in y1, I_s in y2 and I_m in y3. These quantities are given on page B-97 of this *Guide*. Note that M and S contain the unrounded values of the inputs at the critical point.	[calculator screen showing y1=8.61967-1.244S+..., y2=-1.244+.1794S+.2..., y3=...071894M+.2149155]	M 6.89988405123 S -1.33159744075
Take the derivative of y2 = I_s with respect to s and we have I_{ss}, the partial derivative of I with respect to s and then s again. Find $I_{ss} \approx 0.1794$ at the critical point. Take the derivative of y3 = I_m with respect to s and we have I_{ms}, the partial derivative of I with respect to m and then s. Find $I_{ms} \approx 0.2149$ at the critical point.		der1(y2,S,S) .1794 der1(y3,S,S) .214915 If you prefer, enter the numerical value of S in the 3rd position.
Take the derivative of y2 = I_s with respect to m and we have I_{sm}, the partial derivative of I with respect to s and then m. Find $I_{sm} \approx 0.2149$ at the critical point. Take the derivative of y3 = I_m with respect to m and we have I_{mm}, the partial derivative of I with respect to m and then m again. Find $I_{mm} \approx 0.0719$ at the critical point.		der1(y2,M,M) .214915 der1(y3,M,M) .071894 If you prefer, enter the numerical value of M in the 3rd position.
The second partials matrix is $\begin{bmatrix} I_{ss} & I_{sm} \\ I_{ms} & I_{mm} \end{bmatrix}$. Find the value of $D \approx (0.1794)(0.71894) - (0.214915)^2 \approx -0.033$. Because $D < 0$, the Determinant Test tells us that $(s, m, I) \approx (-1.33, 6.90, 8.72)$ is a saddle point.		.1794(.071894)-.21491 5² -.033290673625

NOTE: The values of the second partial derivatives were not very difficult to determine without the calculator in this example. However, with a more complicated function, we strongly suggest using the above methods to provide a check on your analytic work to avoid making simple mistakes. You can also use the function in y1 to check the derivatives in y2 and y3. We use some randomly chosen values of S and M to illustrate this quick check rather than use the values at the critical point, because the critical point values give 0 for y2 and y3.

Store different values in S and M and evaluate y2 at these values. Then evaluate the TI-86's derivative of y1 with respect to S at these same values.

Evaluate y3 at the stored values of S and M. Then evaluate the TI-86's derivative of y1 with respect to M at these same values.

10.3 Optimization Under Constraints

Optimization techniques on your calculator when a constraint is involved are the same as the ones discussed in Sections 10.2.1 and 10.2.2 except that there is one additional equation in the system of equations to be solved.

10.3.1 FINDING OPTIMAL POINTS UNDER CONSTRAINED OPTIMIZATION
You can use any of the methods indicated in Sections 10.2.1 or 10.2.2 of this *Guide* to find critical points under constrained optimization. We choose to illustrate the process using Method 2 for matrices to solve the linear system that results in the sausage cooking illustration that is at the beginning of Section 10.3 of *Calculus Concepts*. The system of equations, written with the constant terms on the right-hand side of the equations and the coefficients of the three input variables in the same positions on the left-hand side of each equation, is

$$-5.83s - \lambda = -11.78$$
$$-5.83w \quad\quad - \lambda = -11.69$$
$$w + s \quad\quad = 1$$

(The order in which the variables appear does not matter as long as it is the same in each equation. Note that we use a different order than in the text.)

When a particular variable does not appear in an equation, its coefficient is zero. Access the simultaneous linear equation solver with [2nd] [TABLE] (SIMULT).

Enter 3 as the number of equations. Enter the coefficients of $x_1 = w$, $x_2 = s$, and $x_3 = \lambda$ in the *a* positions and the constant terms in the *b* positions for the first two equations.	
Repeat the process the third equation. Then press [F5] [SOLVE].	

The solution to the system of equations is $w \approx 0.492$, $s \approx 0.508$, and $\lambda = 8.82$. Then, using the [STO▸] key, store x_1 to W, x_2 to S, and x_3 to L so that the unrounded values will be available for use. See the discussion on pages B-95 and B-96 for more detailed instructions.

10.3.2 CLASSIFYING OPTIMAL POINTS UNDER CONSTRAINED OPTIMIZATION
Classifying critical points when a constraint is involved is done by graphing the constraint on a contour graph or by examining close points. Your TI-86 cannot help with the contour graph classification. We illustrate the procedure used to examine close points for the function given

in Example 1 of Section 10.3 of *Calculus Concepts* – the Cobb-Douglas production function $f(L, K) = 48.1L^{0.6}K^{0.4}$ subject to the constraint $g(L, K) = 8L + K = 98$.

We first find the critical point(s). Because this function does not give a linear system of equations, we use the algebraic method indicated in Section 10.2.2. The system of equations is

$$\begin{cases} 28.86L^{-0.4}K^{0.4} = 8\lambda \\ 19.24L^{0.6}K^{-0.6} = \lambda \\ 8L + K = 98 \end{cases} \Rightarrow \quad 28.86L^{-0.4}K^{0.4} = 8(19.24L^{0.6}K^{-0.6}) \quad [5]$$

$$\Rightarrow \quad K = 98 - 8L \quad [6]$$

Clear the y(x)= list. Enter the function f in y1 and the expression for K in y2. Rewrite the *other* equation (equation 5) so that it equals 0, and enter the non-zero side in y3. Then substitute* y2 into y3.	

*Remember that K = y2. Put the cursor on the first K in y3 and replace K by y2. Note that you will need to use the [2nd] [DEL] (INS) key to type "2" in y2. Do the same for the other K in y3. The expression now in y3 is the left-hand side of an equation that equals 0 and contains only one variable, namely L. (We are not sure how many answers there are to this equation.)

Use the SOLVER to solve the equation y3 = 0. Try several different guesses and see that they all result in the same solution. On the home screen call up y2 to find K. (Store this value in K.) y1 gives the value of f at this point.	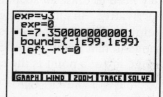

We now need to test close points to see if this output value of f is maximum or minimum. You need to remember that whatever close points you choose, they must be near the critical point and *they must be on the constraint g*.

WARNING: Do not round during this procedure. Rounding of intermediate calculations and/or inputs can give a false result when the close point is very near to the optimal point.

Choose a value of L that is less than $L = 7.35$, say 7.3. Find the value of K so that $8L + K = 98$. Remember that K = y2, so store 7.3 in L and call up y2. Find f at L = 7.3, K = 39.6.
 	
 At this close point, the value of f is less than the value of f at the optimal point.
	7.3→L:y2
	                   39.6
	 Ans→K
	                   39.6
	 y1
	         690.584564125

Choose another value of L, this time one that is more than $L = 7.35$, say 7.4. Find the value of K so that $8L + K = 98$ by storing 7.4 in L and calling up y2. Find f at L = 7.4, K = 38.8.
 	
 At this close point, the value of f is less than the value of f at the critical point. Thus, (7.35, 39.2, 690) is a maximum point.
	690.584564125
	 7.4→L:y2
	                   38.8
	 Ans→K
	                   38.8
	 y1
	         690.584455414

Chapter 11 Dynamics of Change: Differential Equations and Proportionality

The TI-86 offers many opportunities when studying differential equations. It draws slope field graphs and has as built-in commands many of the techniques that must be programmed into other calculators. However, in this *Guide*, we look only at those tools that are necessary for the study of differential equations as presented in Chapter 11 of *Calculus Concepts*. We encourage you to further explore the TI-86 capabilities by reading and working through pages 132-150 of your *TI-86 Graphing Calculator Guidebook*.

11.3 Numerically Estimating by Using Differential Equations: Euler's Method

Many of the differential equations we encounter have solutions that can be found by determining an antiderivative of a given rate-of-change function. Thus, many of the techniques that we learned using the TI-86's numerical integration function apply to this chapter. (See Chapters 6 and 7 of this *Guide*.)

11.3.1 EULER'S METHOD FOR A DIFFERENTIAL EQUATION WITH ONE INPUT VARIABLE
You may encounter a differential equation that cannot be solved by standard methods and you may need to draw an accumulation graph for a differential equation without first finding an antiderivative. In either of these cases, numerically estimating a solution using Euler's method is helpful. This method relies on the use of the derivative of a function to approximate the change in that function. Recall from Section 5.1 of *Calculus Concepts* that the approximate change in f at a point is the rate of change of f at that point times a small change in x. That is,

$$f(x + h) - f(x) \approx f'(x) \cdot h \text{ where } h \text{ represents the small change in } x.$$

Now, if we let $b = x + h$ and $x = a$, the above expression becomes

$$f(b) - f(a) \approx f'(a) \cdot (b - a) \quad \text{or} \quad f(b) \approx f(a) + (b - a) f'(a)$$

The starting values for the coordinates of the point (a, b) will be given to you and are often called the *initial condition*. The next step is to repeatedly apply the formula given above to use the slope of the tangent line at $x = a$ to approximate the change in the function between the inputs a and b. When h, the distance between a and b, is fairly small, Euler's method will often give close numerical estimates of points on the solution to the differential equation containing $f'(x)$.

WARNING: Be wary of the fact that there is some error involved in each step of the Euler approximation process that is compounded when each result is used to obtain the next result.

We illustrate Euler's method for a differential equation containing one input variable with the differential equation in Example 1 of Section 11.3. This equation gives the rate of change of the total sales of a computer product t years after the product was introduced:

$$\frac{dS}{dt} = \frac{6.544}{\ln(t + 1.2)} \text{ billion dollars per year}$$

Because Euler's method involves a repetitive process, a program that performs the calculations used to find the approximate change in the function can save you time and eliminate computational errors and some error in rounding.

TI-86 Guide

B-103

Before using this program, you must have the differential equation in the y1 location of the y(x)= list with x as the input variable. From the home screen, access the Euler program with the keys [PRGM] [F1] [NAMES].	`\y1■6.544/ln (x+1.2)`

The code for program EULR is listed in the *TI-86 Program Appendix*. Run the program. Each time the program stops for input or for you to view a result, press [ENTER] to continue.

We choose to use 16 steps. Enter this value. The interval is 4 years, so enter the step size = $\frac{\text{length of interval}}{\text{number of steps}} = \frac{4}{16} = 0.25$.	HAVE dy/dx in y1 Number of steps= 16 Step size= 0.25	
The initial condition is given as the point (1, 53.2). Enter these values when prompted for them.	Number of steps= 16 Step size= 0.25 Initial input= 1 Initial output= 53.2 Input,Output is 1.25 55.2749370245	The first application of the formula gives an estimate for the value of the total sales at $x = 1.25$: $S(1.25) \approx 55.275$
Press [ENTER] several more times to obtain more estimates for total sales. Record the input values and output estimates on paper as the program displays them.	57.1006513204 Input,Output is 1.75 58.747766427 Input,Output is 2 60.2600535235	Continue pressing the [ENTER] key to obtain more estimates of points on the total sales function S.
When 16 steps have been completed (that is, after the input reaches 5), the program draws a graph of the points *(input, output estimate)* connected with straight line segments. This graph is an estimate of the graph of the differential equation solution.	71.7020961191 Input,Output is 4.75 72.6420741685 Input,Output is 5 73.5594275692	This is an estimate of the graph of the function $S(t)$.
In the y(x)= list, turn Plot3 off, turn y1 on, and press [EXIT] [F3] [ZOOM] [MORE] [F1] [ZFIT] to draw the graph of the differential equation.	`\y1■6.544/ln (x+1.2)`	This is the slope graph – the graph of $S'(t)$.

11.3.2 EULER'S METHOD FOR A DIFFERENTIAL EQUATION WITH TWO INPUT VARIABLES

Program EULR can be used when the differential equation is a function of x and y with $y = f(x)$. Follow the same process that is illustrated in the previous section of this *Guide*, but enter $\frac{dy}{dx}$ in y1 using the letters x and y as they are written in the given equation.

If the differential equation is written in terms of variables other than x and y, let the derivative symbol be your guide as to which variable corresponds to the input and which corresponds to the output. For instance, if the rate of change of a quantity is given by $\frac{dP}{dn} = 1.346P(1 - n^2)$, compare $\frac{dy}{dx}$ to $\frac{dP}{dn}$, entering in y1 the expression $1.346y(1 - x^2)$.

Copyright © Houghton Mifflin Company. All rights reserved.

The differential equation may be given in terms of y only. For instance, if $\frac{dy}{dx} = k(30-y)$ where k is a constant, enter y1 = K(30 - y). Of course, you need to store a value for k or substitute a value for k in the differential equation before using program EULR. It is always better to store the exact value for a constant instead of using a rounded value.

WARNING: When using program EULR, you must use a lower-case y when entering an equation involving two variables in the y(x)= list. Use 2nd ALPHA (alpha) 0 (y) to type y.

We illustrate using Euler's method with two input variables by using the equation in Example 2 of Section 11.3 in *Calculus Concepts*.

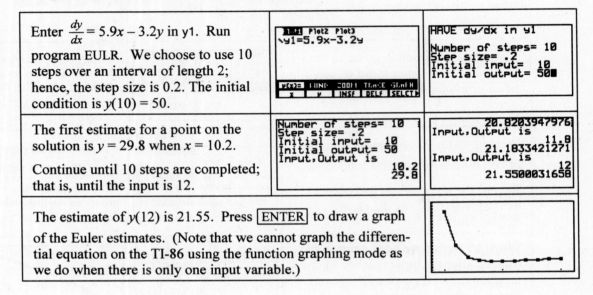

Enter $\frac{dy}{dx} = 5.9x - 3.2y$ in y1. Run program EULR. We choose to use 10 steps over an interval of length 2; hence, the step size is 0.2. The initial condition is $y(10) = 50$.
The first estimate for a point on the solution is $y = 29.8$ when $x = 10.2$. Continue until 10 steps are completed; that is, until the input is 12.
The estimate of $y(12)$ is 21.55. Press ENTER to draw a graph of the Euler estimates. (Note that we cannot graph the differential equation on the TI-86 using the function graphing mode as we do when there is only one input variable.)

Appendix: Trigonometry Basics

Before you begin working through the ideas in this *Appendix*, be certain that the MODE screen in the basic setup has Radian/Degree set to Radian measure.

A.1   FINDING OUTPUTS OF TRIG FUNCTIONS WITH DEGREE INPUTS   It is very important that you have the correct mode set when evaluating trigonometric function outputs. Unless you see the degree symbol in a problem, the mode setting should be Radian. We show two different methods of evaluating trig functions with degree inputs, using the illustration as given in the discussion following Figure A9 in the *Trigonometry Appendix*.

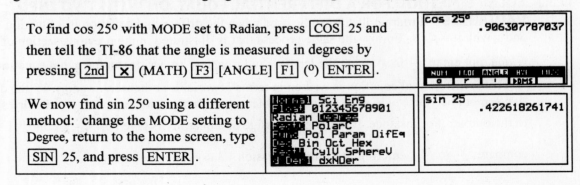

To find cos 25° with MODE set to Radian, press COS 25 and then tell the TI-86 that the angle is measured in degrees by pressing 2nd × (MATH) F3 [ANGLE] F1 (°) ENTER.
We now find sin 25° using a different method: change the MODE setting to Degree, return to the home screen, type SIN 25, and press ENTER.

WARNING: Do not forget to change the MODE setting back to Radian if you use this second method. If not, you will obtain an incorrect value the next time you evaluate a trig function for an angle measured in radians or when drawing a graph. We recommend using the first method.

A.2 CONVERTING ANGLES FROM RADIANS TO DEGREES

Even though the formula for this conversion is simple, the TI-86 converts angles from radians to degrees quite easily. We illustrate using the angle in part *a* of Example 7 in the *Trigonometry Appendix*.

As shown in Example 7, convert $7\pi/6$ radians to degrees with multiplication by the proper factor. Let the units be your guide and carefully use parentheses: $\left(\dfrac{7\pi}{6} \text{ radians}\right)\left(\dfrac{180°}{\pi \text{ radians}}\right) = 210°$

Note that the MODE setting does not matter because you are not using the ANGLE menu or any of the trig function keys. Next, we see how to have the TI-86 do the conversion.

Set MODE to Degree. Type $7\pi/6$ and press [2nd] [×] (MATH) [F3] [ANGLE] [F2] (ʳ) [ENTER]. You must put $7\pi/6$ in parentheses or the calculator assumes only the 6 is measured in radians.

NOTE: When MODE is set to Degree, the TI-86 assumes all angles (inputs and outputs) are measured in degrees. The radian setting (ʳ) in the ANGLE menu overrides the MODE setting.

A.3 CONVERTING ANGLES FROM DEGREES TO RADIANS

We illustrate converting an angle from degree measure to radian measure using the angle in part *b* of Example 7 in the *Trigonometry Appendix*. As indicated in the last section, you can convert using the fact that π radians = 180 degrees or let the TI-86 do the conversion.

As shown in Example 7, convert 135 degrees with multiplication by the proper factor. Let the units be your guide and carefully use parentheses: $(135°)\left(\dfrac{\pi \text{ radians}}{180°}\right) = \dfrac{3\pi}{4} \approx 2.36$ radians.

Note that the MODE setting does not matter because you are not using the ANGLE menu or any of the trig function keys.

(See page B-2 for entering the fraction conversion key.)

To have the TI-86 do the conversion, set MODE to Radian, type 135, and press [2nd] [×] (MATH) [F3] [ANGLE] [F1] (°). Press [ENTER].

NOTE: When MODE is set to Radian, the TI-86 assumes all angles (inputs and outputs) are measured in radians. The degree choice (°) in the ANGLE menu overrides the MODE setting.

TroubleShooting the TI-86

❑ **CALCULATOR BATTERIES**

See the information in CALCULATOR BATTERIES in *TroubleShooting the TI-83* on page *A*-107 of this *Guide*.

❑ **DELETING PROGRAMS OR OTHER INFORMATION FROM THE TI-86**

- *How do I delete information from my calculator?*

Press **2nd 3 [MEM] F2 [Delete]** and then press the F-key corresponding to the type of item you are deleting. For instance, to delete a program, press **MORE F5 [Prgm]**. Use the down arrow to move the cursor next to the information you want to delete and press **ENTER**. **EXIT** returns you to the home screen.

- *When I try to run a program, I keep getting ERROR 10 DATA TYPE. What can I do?*

You probably stored a picture or graph data base to a variable name that conflicts with a variable name used by the program. Press **2nd 3 [MEM] F2 [Delete] F1 [ALL]**, scroll down the list of variables and see if you have one that says GDB (graph data base) or PIC (picture). If you find one, and it is not something you want to keep, have the cursor next to the name and delete it by pressing **ENTER**. (Be very careful on this screen – pressing **ENTER** deletes what-ever the cursor is next to and there is no undo instruction!)

If the **PIC** or **GDB** is something you do want to save, return to the home screen with **EXIT**, press **GRAPH MORE RECDB** (recall graph data base) or **GRAPH MORE MORE RCPIC** (recall picture) and type in the current name of your picture or graph data base, press **ENTER** and then either **STPIC** or **STGDB** (you may have to press **MORE** to find these). Type in the new name and press **ENTER**. Be sure that the new name has more than a single letter! Return to the delete screen (see the first paragraph) and delete the old graph data base or picture. This should correct the problem.

❑ **GENERAL INFORMATION**

- *Why do my calculator screens not look like the ones in this Guide?*

Make sure your calculator settings are as specified in the Setup Instructions on page *B*-1.

- *How do I make the display screen darker or lighter?*

See the response to this question on page *A*-107.

- *I get tired of pressing so many keys to have to find instructions that I use all the time. Is there something I can do to make the process simpler?*

Yes, you can access the instructions in the catalog (See page *B*-40 of this *Guide*) or the TI-86 has a CUSTOM menu in which you can enter commands that you use many times. See pages 44-45 of the *TI-86 Graphing Calculator Guidebook*.

- *Is there a difference in lower-and upper-case letters on the TI-86?*

Yes, the TI-86 makes a distinction in the case. For instance, *P* and *p* are different variable names on this calculator. Also, the TI-86 considers *PT* as the name of a single variable, but *P*T* or *P*(*T*) means the multiplication of *P* and *T*.

- *What causes an ERROR 14 UNDEFINED message?*

Anytime you use a variable consisting of one or more letters, it must first be defined by storing a value to that variable. For example, to use *P* in a formula, you must first store some value to *P*. Another time that this error occurs (especially in Chapters 9 and 10 of *Calculus Concepts*) is when you forget to put a times sign in between two letters that each stand for different variables. For instance, if you enter in the graphing list (or the solver) the equation y1 = 2A + B^2 + 3AB, the TI-

86 thinks that there are three inputs: A, B, and AB. This equation should be entered on one line as $y1 = 2A + B^2 + 3A*B$.

- *I have either lost or never got the owner's manual for the TI-86. What can I do?*

The *TI-86 Graphing Calculator Guidebook* is available at the Texas Instruments web site with address **www.TI.com/calculators**.

❏ GRAPHING

- *Why don't I see a graph when I press the correct keys to make it draw?*

Make sure that you have the function entered in the y(x)= (graphing) list, using a *lower-case x* for the input variable. See the remainder of the response to this question on page *A*-108.

- *I've lost y1 (and/or y2, etc.) in the graphing list. How do I get them back?*

If you don't need any of the current functions in the list, press **DELf** on the graphing menu until it resets to y1. If you want to save the functions that are there, place the cursor on the line above which you want to insert a function location. Press **INSf** on the graphing menu.

- *While trying to draw the graph of an equation that I entered in the y(x)= list, I get an ERROR 13 DIMENSION. What do I do?*

The TI-86 will not draw the graph of a function when one of the scatter plots is turned on and the lists are of unequal length or contain no data. To correct this problem, choose **1: [Quit]** and turn off all stat plots by making sure that the names (Plot 1, Plot 2, Plot 3) are not dark. The graph should now draw with out any problems.

- *Do I have to clear the function location in the y(x)= list before pasting in another function when finding an equation of best-fit?*

See the response to this question on page *A*-108.

❏ LISTS

- *I don't have all the lists when I press **2nd + (STAT) F2 [EDIT]**. How do I fix this?*

See the response to this question on page *A*-108.

❏ PROGRAMS

- *Where can I get the programs that are used in this Guide?*

See the response to this question on page *A*-108.

- *Why won't a program run, or why do I get an error when I try to run a program?*

You may have unknowingly deleted or altered one or more lines of the program code. The easiest method of fixing the problem is to use the link chord to re-transfer the entire program into your calculator. If this is not possible, you can find the correct program code in the *TI-86 Program Appendix* at the *Calculus Concepts* Website. Press **PRGM right arrow [EDIT]** and **down arrow** until the name of the corrupt program is highlighted and press **ENTER**. Compare the code in your calculator with what is printed in the *TI-86 Program Appendix* until you find the error and correct it by retyping the correct code. Consult the *TI-86 Graphing Calculator Guidebook* that came with your calculator for the location of the symbols in the program code.

- *Program NUMINTGL keeps giving an error when I try to run it. What can I do?*

Delete all functions from the y(x)= list (except the function in y1) before using the program named NUMINTGL. This program uses every memory location in the calculator and all the variables have to be defined before the program will run. (See the fifth bulleted item under GENERAL INFORMATION. Also see DELETING PROGRAMS OR OTHER INFORMATION FROM THE TI-86.)